T0092032

Of Microbes and Art

Of Microbes and Art

The Role of Microbial Communities in the Degradation and Protection of Cultural Heritage

Edited by

Orio Ciferri
University of Pavia
Pavia, Italy

Piero Tiano
C.N.R.–C.S. "Opere d'Arte"
Florence, Italy

and

Giorgio Mastromei
University of Florence
Florence, Italy

Springer Science+Business Media, LLC

Library of Congress Cataloging-in-Publication Data

Of microbes and art: the role of microbial communities in the degradation and protection of cultural heritage/edited by Orio Ciferri, Piero Tiano, and Giorgio Mastromei
p. cm.
Proceedings of an International Conference on Microbiology and Conservation (ICMC) held June 17–19 in Florence, Italy.
Includes bibliographical references and index.
ISBN 978-0-306-46377-8 ISBN 978-1-4615-4239-1 (eBook)
DOI 10.1007/978-1-4615-4239-1
1. Biodegradation—Congresses. 2. Art objects-Conservation and restoration—Congresses. 3. Materials—Biodeterioration—Congresses. I. Ciferri, Orio, 1928– II. Tiano, Piero, 1947– III. Mastromei, Giorgio, 1951– IV. International Conference on Microbiology (Florence, Italy)

QR135 .O36 2000
701'.8'8—dc21

00-023508

ISBN 978-0-306-46377-8

Proceedings of an International Conference on Microbiology and Conservation (ICMC) entitled Of Microbes and Art: The Role of Microbial Communities on the Degradation and Protection of Cultural Heritage, held 17–19 June, 1999, in Florence, Italy.

Originally published by Kluwer Academic/Plenum Publishers in 2000

International Scientific Committee

Contributors

Abbruscato Pamela. *Department of Food and Microbiological Sciences and Technologies, University of Milan, Italy.*
Abdulla Hesham. *Department of Botany, Suez Canal University, Isamailia, Egypt.*
Biagiotti Lucia. *Department of Animal Biology and Genetics "Leo Pardi", University of Florence* and *Center for Scientific Conservation of Art Works, CNR, Florence, Italy.*
Borin Sara. *Department of Food and Microbiological Sciences and Technologies, University of Milan, Italy.*
Bracci Susanna. *Center for Scientific Conservation of Art Works, CNR, Florence, Italy.*
Castanier Sabine. *Laboratory of Biogeology and Microbiogeology, University of Nantes, France.*
Ciferri Orio. *Department of Genetics and Microbiology "A. Buzzati Traverso", University of Pavia, Italy.*
Daffonchio Daniele. *Department of Food and Microbiological Sciences and Technologies, University of Milan, Italy.*
Daly Simona. *Department of Animal Biology and Genetics "Leo Pardi", University of Florence, Italy.*
De Hoog Sybren. *Centraalbureau voor Schimmelcultures, Baarn, The Netherlands.*
De Leo Filomena. *Institute of Microbiology, University of Messina, Italy.*
Dewedar Ahmed. *Department of Botany, Suez Canal University, Ismailia, Egypt.*
Dornieden Thomas. *ICBM, University "Carl von Ossietzky", Oldenburg, Germany.*

Florian Mary-Lou Esther. *Royal British Columbia Museum, Victoria, Canada.*
Galizzi Alessandro. *Department of Genetics and Microbiology "A. Buzzati-Traverso", University of Pavia, Italy.*
Gorbushina Anna A. *ICBM, University "Carl von Ossietzky", Oldenburg, Germany.*
Gurtner Claudia. *Institute of Microbiology and Genetics, University of Vienna, Austria.*
Hermosin Bernardo *Instituto de Recursos Naturales y Agrobiologia, C.S.I.C., Seville, Spain.*
Koestler Robert J. *Sherman Fairchild Center for Objects Conservation, The Metropolitan Museum of Art, New York, USA.*
Krumbein Wolfgang E. *Geomicrobiology, ICBM, University "Carl von Ossietzky", Oldenburg, Germany.*
Laiz Leonilla *Instituto de Recursos Naturales y Agrobiologia, C.S.I.C., Seville, Spain.*
Lamenti Gioia. *Center for Study of Autotrophic Microorganisms, CNR, Florence, Italy.*
Le Métayer-Levrel Gaële. *Laboratory of Biogeology and Microbiogeology, University of Nantes, France.*
Loubière Jean-François. *CALCITE S.A., Levallois-Perret, France.*
Lubitz Werner. *Institute of Microbiology and Genetics, University of Vienna, Austria.*
Maifreni Tullia. *Stazione Sperimentale per la Seta, Milan, Italy.*
Mastromei Giorgio. *Department of Animal Biology and Genetics "Leo Pardi", University of Florence, Italy.*
Matteini Mauro. *Opificio Pietre Dure, Florence, Italy.*
May Eric. *School of Biological Sciences, University of Portsmouth, UK.*
Orial Geneviève. *Laboratoire de Recherche des Monuments Historiques, Champs sur Marne, France.*
Papida Sophia. *School of Biological Sciences, University of Portsmouth, UK.*
Perito Brunella. *Department of Animal Biology and Genetics "Leo Pardi", University of Florence, Italy.*
Perthuisot Jean-Pierre. *Laboratory of Biogeology and Microbiogeology, University of Nantes, France.*
Pinar Guadalupe. *Institute of Microbiology and Genetics, University of Vienna, Austria.*
Ranalli Giancarlo. *Department of Food and Microbiological Sciences and Technologies, University of Molise, Campobasso, Italy.*
Realini Marco *Gino Bozza Center, CNR, Milan, Italy.*
Recio Delfina *Instituto de Recursos Naturales y Agrobiologia, C.S.I.C., Seville, Spain.*

Rölleke Sabine. *Institute of Microbiology and Genetics, University of Vienna, Austria.*
Romanò Maria. *Stazione Sperimentale per la Seta, Milan, Italy.*
Saiz-Jimenez Cesareo. *Instituto de Recursos Naturales y Agrobiologia, C.S.I.C., Seville, Spain.*
Salvadori Ornella. *Soprintendenza per i Beni Artistici e Storici di Venezia, Venice, Italy.*
Scicolone Giovanna. *Scuola Regionale per la Conservazione dei Beni Culturali, Botticino Sera (BS), Italy.*
Seves Alberto. *Stazione Sperimentale per la Cellulosa, Carta e Fibre Tessili Vegetali ed Artificiali, Milan, Italy.*
Seves AnnaMaria. *Department of Genetics and Microbiology "A. Buzzati Traverso", University of Pavia, Italy.*
Sora Silvio. *Department of Genetics and Microbiology "A. Buzzati Traverso", University of Pavia, Italy.*
Sorlini Claudia. *Department of Food and Microbiological Sciences and Technologies, University of Milan, Italy.*
Sterflinger Katja. *Geomicrobiology, ICBM, University "Carl von Ossietzky", Oldenburg, Germany.*
Tayler Sally. *School of Biological Sciences, University of Portsmouth, UK.*
Tiano Piero. *Center for Scientific Conservation of Art Works, CNR, Florence, Italy.*
Tomaselli Luisa. *Center for Study of Autotrophic Microorganisms, CNR, Florence, Italy.*
Tosini Isetta. *Opificio Pietre Dure, Florence, Italy.*
Urzì Clara. *Institute of Microbiology, University of Messina, Italy.*
Warscheid Thomas. *Institute for Material Science, Bremen, Germany.*
Zanardini Elisabetta. *Department of Food and Microbiological Sciences and Technologies, University of Milan, Italy.*

Preface

Cultural heritage, the term now utilized to cover the immensely diverse mass of documents of all types upon which our societies confer a particular artistic, historic or ethnologic value, provides an extremely wide range of ecological niches and chemical compounds which may be exploited by an equally wide range of microorganisms. As a consequence, microorganisms colonize all types of cultural artifacts (from archaeological sites to miniatures, from illuminated parchment to stone monuments) often causing extensive and irreversible aesthetic and structural damages. Further, biodeterioration induced by microorganisms is a global problem since microorganisms are present in all habitats, including the most extreme ones, and possess an amazingly diversified metabolic versatility. Therefore, microbial defacement and degradation of cultural artifacts, the traces of human civilisation, is a worldwide problem: all countries are affected regardless of their history, geographical localization, socio-economic conditions, etc. There is an allegedly Italian saying that sums this up by stating "Blessed were the ancients, for they had no antiquities".

Already in the thirties reports linking defacement of cultural heritage to colonization by microorganisms appeared and attempts were made to control such "infestations" by mechanical, chemical or physical methods. However, only in the fifties did special sessions devoted to the role that microbial colonization could play in the defacement and degradation of cultural heritage began to be organized in the framework of more general meetings dedicated to the studies on conservation. This led to the realization that biodeterioration could also be caused by the interactions of microbes with "cultural substrates", the term substrate meaning the surfaces on which a microorganism may settle and grow as well as the sources of energy and chemicals necessary for its growth. This book is the outcome of a meeting, held in Florence in 1999, dedicated exclusively to the relationships existing between microorganisms and cultural heritage. It seemed to the organizers that, although enough data is now available demonstrating a direct link between colonization by microbes and defacement of cultural artifacts, a frustrating ignorance concerning the mechanisms of such interactions continues to exist. The meeting was intended to address these issues but also to gaze at the future developments in these areas.

Personally, I have found extremely gratifying that almost one hundred scientists, many from overseas, attended the Florence conference and that the greatest majority (over sixty percent) of the registered participants came from outside Italy. It was even more gratifying to realize that many were young scientists (postdoctorate fellows, graduate students and the like) indicating that, perhaps, there is a younger generation who is concerned by the biodegradation of cultural heritage and considers the possibility of working in these areas, conjugating laboratory research and conservation, scientific quest and practical applications. I also received the impression that this meeting has represented an important step in the study of the relations between microbiology and conservation. It seemed to me that many of the communications presented at the conference witnessed the coming of age of microbiological research in the fields of conservation. Just to cite a few examples, the utilization of more sophisticated and more sensitive analytical techniques such as PCR amplification and molecular fingerprinting for the identification of the microbial flora present on cultural artifacts, the elaboration of laboratory models that allow to study microbially-induced biodegradation under controlled conditions and in much shorter periods of time, the possibility of a sort of microbial bioremediation of stone artifacts, the demonstration that microbial colonization is almost always accomplished through the formation of biofilms on the surface of the cultural objects.

It emerged also from the meeting that a lot of work remains to be done. If we have made a quantum leap in the techniques for identifying the microbial taxa present on cultural heritage (I look with amazement and, at the same time, with awe at the lengthening of the lists of microorganisms encountered on art objects), we know very little about, for instance, the species responsible for the degradation process, how the flora has evolved, the interactions among the different members of the microbial biodegradative communities, or the mechanisms underlying the chemical and mechanical transformations brought about by the action of microorganisms. Borrowing from the terminology of medical microbiology, one could say that very often the causative agent has not been identified and little is known concerning the microbe-host interactions. The study of microbial colonization of cultural heritage is difficult, as it deals with complex systems and their interrelationships: the materials composing the object, the environment in which the object is localized, the microbial flora surrounding and colonizing the object. Further, with few exceptions, colonization is due to a community of different microorganisms interacting with the substrate but also among themselves.

Another message I received at the Florence meeting is the need for a closer collaboration between experimental scientists on one hand and conservators on the other. It is possible that two distinct cultures exist here too. On one side the laboratory scientist, whose main interest is to ascertain how a given artifact has changed over the time, which organisms are present on it and how these organisms have contributed to the variations in the physico-chemical properties of the artifact. On the other side, there is the conservator whose main purpose is not so much to understand what has happened and why, but what to do in order to arrest or prevent further injuries to the object. This gap may be bridged only if the two communities get closer and learn to communicate one another. Better coordination and communication are the greatest challenges facing the two communities. My wish is that the future conferences on microbiology and conservation will see the bridging of this gap. Otherwise, there is the risk that, even for cultural heritage, Pasteur's admonition "Messieurs, c'est les microbes qui auront le dernier mot" may come true.....

Orio Ciferri

Chairman of the
Scientific Committee

Contents

Of Microbes and Art

PART 1

ECOLOGY OF MICROBIAL COMMUNITIES DEVELOPING ON ART WORKS

RECENT ADVANCES IN THE MOLECULAR BIOLOGY AND ECOPHYSIOLOGY OF MERISTEMATIC STONE-INHABITING FUNGI

Clara Urzì[1], Filomena De Leo[1], Sybren De Hoog[2] and Katja Sterflinger[3]
[1]Institute of Microbiology, University of Messina, Salita Sperone 31, I-98166 Villaggio S. Agata, Messina, Italy; [2]Centraalbureau voor Schimmelcultures, P.O. Box 273, NL3740 AG Baarn, The Netherlands; [3]Geomikrobiologie ICBM, Carl von Ossietzky Universität Oldenburg, P.O. Box 2503, D-26111 Oldenburg, Germany.

Key words: black fungi, biodeterioration, taxonomy, ecology

Abstract: Fungi are among the most efficient groups of microorganisms causing biodeterioration of organic and inorganic materials. They are remarkably versatile in surviving in unsuitable environmental conditions. Climatic factors on rock and stone monuments in dry subtropical areas may be too extreme to support growth of lichens; nevertheless they allow the colonisation by some groups of well-adapted extremophilic fungi. Over the last years isolation techniques have been improved and have revealed the ubiquitous presence of a diversity of slow growing fungi. A particular role is assumed for dematiaceous meristematic fungi. These fungi have been located especially on rock and stone monuments in the Mediterranean Basin, but it seems that their presence is widespread over different climatic conditions. Studies applied to microbial ecology aiming to establish the involvement of fungi in the decay of stone monuments are usually carried out. Isolation of strains and determination of their phenotypic and genotypic characteristics (morphology, development, metabolism, physiology and phylogeny) allow to distinguish the species from superficially unrelated organisms and to define the ecological role of individual species. In addition, the definition of range of growth and the responses to abiotic factors provide auto-ecological information on their habitats. Genotypic characterization gives a better definition of the studied taxa.

1. INTRODUCTION

Black or Dematiaceous fungi are the most harmful microorganisms associated with stone and monument biodeterioration.

They are commonly isolated from the sun-exposed surfaces in the Mediterranean as well as from dry and cold climates, where they are associated to biological alterations (Anagnostidis et al., 1992; Braams, 1992; Garcia-Valles et al., 1997; Hirsch et al., 1995; Sterflinger, 1995; Urzì et al., 1998).

The term meristematic fungi defines the specific pattern of growth of these fungi especially when they grow in a stressed or extreme environment. Meristematic growth (that can be the sole known modality of growth observed for most of black fungi) is characterised by the production of swollen isodiametric cells with thick cell walls, in which melanin (mostly DHN type) is usually deposited. They remain metabolically active for long periods of time even if low amounts of nutrients are available. Endoconidiation is considered to be favourable to the organism because local dissolution of protective cell walls during germination or budding remains absent and hence the cells are less vulnerable to environmental stress (Wollenzien et al., 1995).

On rock surfaces as well as on the monuments from which black meristematic fungi were isolated, their direct evidentiation on the stone showed their characteristic pattern of growth or they were visible as clusters of cells (Plate 1a, b). As reported by Sterflinger and Krumbein (1995), this modality of growth is a common feature for most of prokaryotic and eukaryotic stone-inhabitant organisms. Thus, the well known ecological principle "of uniformity" can be applied to those organisms living in a habitat subjected to multiple stresses (Sterflinger and Krumbein, 1995; Sterflinger, 1998).

Morphologically similar fungi were isolated from some small granite flakes sampled in a garden in Switzerland and studied by Turian already in 1975 (Turian, 1975).

Since the fungi isolated from stone and monuments surfaces present characteristics comparable to those of fungi isolated from deserts rocks described by Staley et al. (1982) also the term of "microcolonial fungi" (MCF) was applied to them. Another common term, is "Black yeast", used to describe black fungi that have yeast-like stages of reproduction as well as a meristematic pattern of growth.

Common characteristics of black meristematic fungi, as reported by several authors (Staley et al., 1982; Wheleer and Bell, 1988; Urzì et al., 1992a; Sterflinger, 1998; De Leo et al., 1999), are:

- to form black small cauliflower-like colonies (60-100 μm diam.) directly in the rock (Plate 1c);

- to grow very slowly in normal cultural media (Plate 1d);
- to reproduce by isodiametric enlargement with subdividing cells (Plate 1e);
- to grow in oligotrophic conditions;
- to grow at optimal temperature of 20-25°C and to have high resistance to desiccation;
- to resist to osmotic stress and some of them are even able to grow in the presence of high concentration of salts;
- to resist to high and long duration UV radiations (Fig. 1).

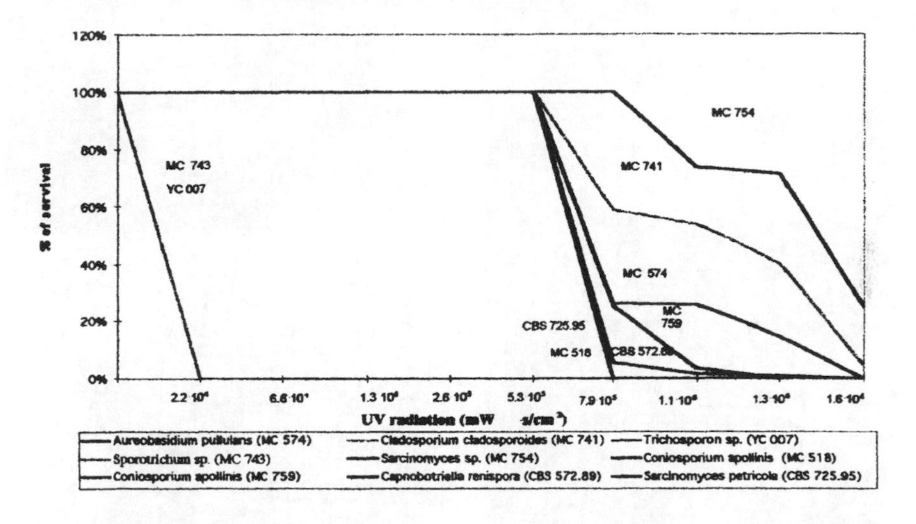

Figure 1. Percentage of survival of black fungi compared with unpigmented strains at different doses of UV radiations.

It has been known for years that black meristematic fungi growing in natural habitats are able to colonise natural outcrops (granite, limestone, calicite, etc.), however they have been only recently "re-discovered" on monuments and extensively studied (Urzì et al., 1992a; Wollenzien et al., 1995; Sterflinger, 1995).

The successful approaches that allowed their optimal isolation can be summarised as follows:

- by using different sampling procedures to isolate these fungi in correspondence of visible (under naked eye or stereomicroscope) black spots (Wollenzien et al., 1995);
- by using media that reduce growth of fast growing and spreading fungi, thus allowing to increase the incubation period (Urzì et al., 1992b).

Sterflinger (1998) showed that MCF are especially adapted to grow on surfaces that are characterised by changes between two environmental situations: high temperatures combined with a low water availability and moderate temperatures combined with water availability. Desiccation seems to be a key factor to resist to high temperatures.

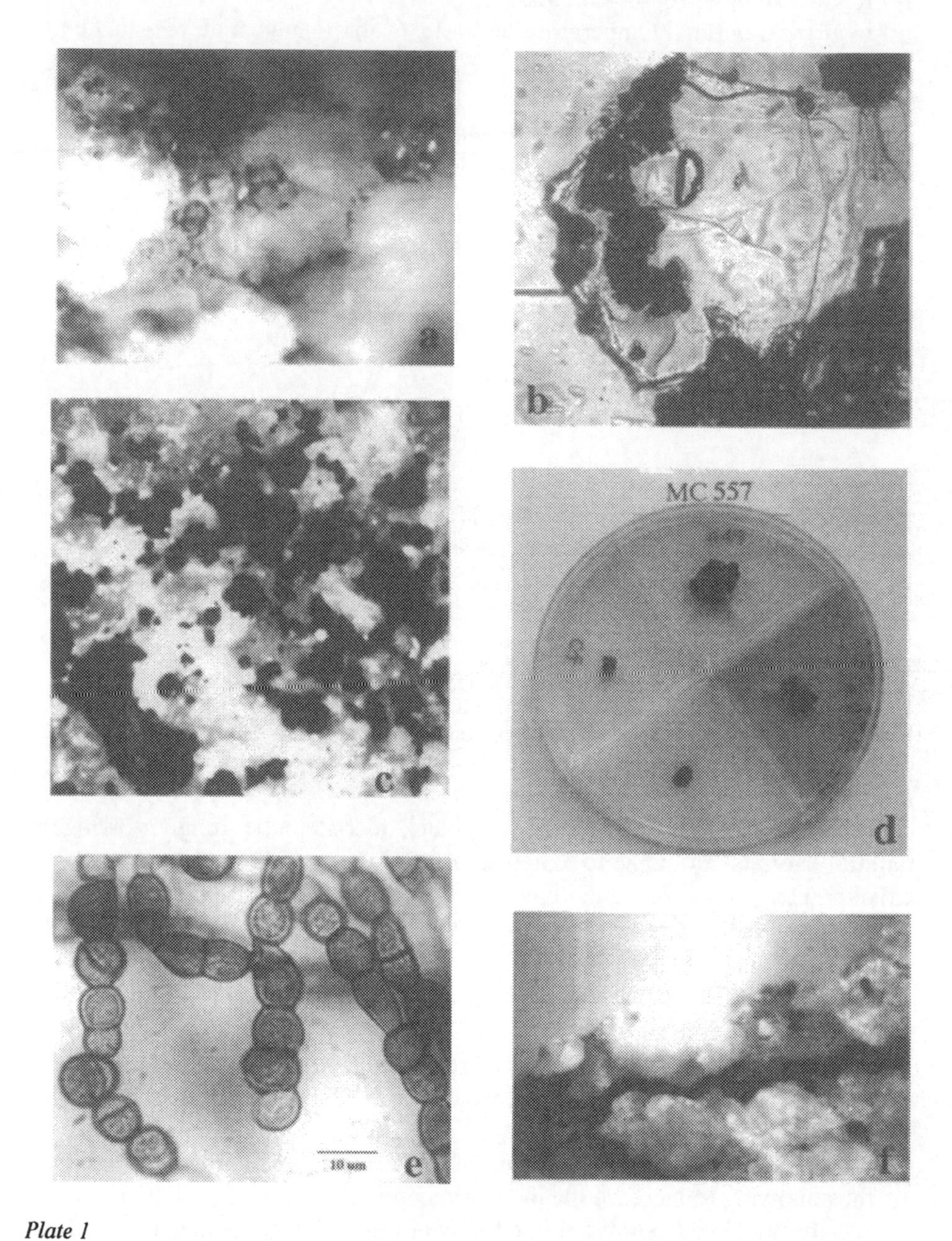

Plate 1

a. Proserpina Sarcophagus (Messina Museum). Thin section of Thassos marble. Chains of black fungi inside the marble. Magnification 32x.

b. Cluster of black fungi closely connected with a marble flake after scotch sampling. Magnification 250x.
c. Black cauliflower-like colonies on Carrara marble surface. Magnification 50x
d. Growth of strain *C. uncinatum* MC557=CBS100212 after one month of incubation at 25°C in 4 different cultural media. Media tested are clockwise: PDA(Top), MEA, OA,CzA.
e. *C. apollinis* MC 525=CBS100214. Isodiametrically dividing cells. Magnification 1000x.
f. Carrara marble section from a sarcophagus located in the garden of Messina Museum. Fungal colonization on cracks. Magnification 20x

Rock-inhabiting black fungi are very difficult to relate with species isolated from plants, soil or other habitats and some species were found only on rocks (Wollenzien et al., 1997; Sterflinger et al., 1997; De Leo et al., 1999).

However, it can be hypothesized that different reservoirs are possible for some if not all the black meristematic fungi living on the stone surfaces. This fact is particularly true for fungi found on the surface (epilithic ones), while for the crypto- and/or chasmo-endolithic fungi, considered as true rock-inhabiting fungi, this is not documented.

Black fungi coming from different hosts and/or substrata can spread and disseminate themselves with different modalities and thus reach stone surfaces. Dissemination through airborne conidia or "fungal units" should be the most diffuse modality of dispersion due to their intrinsic characteristics. In fact, as demonstrated for other black fungi (Yang and Johanning, 1997), they are resistant to desiccation and UV radiation and can present an hydrophobic surface for the easy dispersion by the wind. However, in normal surveys of airborne fungi connected with marble colonisation, the most common findings are those of different genera of *Hyphomycetes* like *Cladosporium* and *Alternaria*, but meristematic fungi have never been isolated from air (Urzì et al., 1999b).

1.1 Bioreceptivity and specificity of rock substrate

It is well known that the nature of the surface that is to be colonised affects the first steps of the process itself. Polished surfaces cannot be easily attacked and thus airborne fungi fail to settle on them (Urzì et al., 1999b).

Fungal biodiversity of colonised surfaces can vary with different rock types as well as in dependence of climatic factors. Fungal density on and within the colonised surfaces is variable depending on the state of conservation of the rocks: crevices and fissures are widely colonised, in alternative pits are the favoured place; in addition, under optimal environmental conditions, the whole surface can also be entirely covered by the colonies. (Plate 2a).

Meristematic black fungi could not be the solitary colonisers of several monuments and natural rock surfaces in mild and humid climate and in protected areas (Braams, 1992; Anagnostidis et al., 1992; Hirsch et al., 1995; Urzì et al., 1998). On the contrary, most rock surfaces in desert climate support the growth of microcolonial structures in the absence of lichens and of cryptoendolithic algae (Staley et al., 1982).

On rocks and monuments exposed to direct sun irradiation and in dry climate, the isolation of MCF is almost exclusive (Sterflinger, 1995), but it was observed that, in some cases, groups of orange to black pigmented bacteria can be found sharing the same habitat if not the same niche (Zezza et al., 1995; Eppard et al., 1996).

Fungal processes of colonisation and further spreading can be studied through fractal geometry (Urzì et al., 1996). The anisometry of the substrate (crystals boundaries, preexisting cracks or cavities) as well as nutrients' availability and climatic parameters control the pattern of colonisation. Thus, fungal spreading on the surface can be foreseen.

Sterflinger and Krumbein (1997) described several steps of colonisation and following biodeteriorative effects:

a) prepenetration - is the moment in which the fungus reaches the rock surface and it settles better in the contacts between crystals (weakest zone). In this stage the fungus produce organs for attachments (Plate 2d, e). This step is strictly connected with the bioreceptivity as described by Guillitte (1995) being an essential prerequisite of the stonc itself to be colonized by any microorganism (Urzì and Realini, 1996);
b) penetration stage, in this phase, the fungus produces and extends its hyphae or chain of cells into the inner part of the rock (Plate 2a);
c) post-penetration stage, in which, once penetrating or forming holes, cavities etc., the fungus is able to produce larger colonies and thus sometimes its development is larger in depth than in the surface (Plate 1f).

Due to the fact that no acid production was demonstrated by these fungi, the mechanical destructive force on marble crystals seems to explain better the process (Krumbein and Urzì, 1993; Urzì and Krumbein, 1994).

1.2 Biodeterioration patterns

Most of the rock microbial colonisers, including black meristematic fungi, are metabolically active for most of part of their life cycles on the substrata. Staley et al. (1982) demonstrated that MCF in the desert rocks were able to colonise newly exposed surfaces after 45 months.

Vitality of colonising microorganisms can also be measured with a chromometer. Shifts in the colour (orange, grey to black) and in its intensity are expression of the presence of microflora, of the seasonal cycles of microorganisms as well as of the diffusion on the surface (Urzì and Realini, 1996, 1998). The alternation of active and stationary cycles of life affects enormously their behaviour on the monuments surfaces, provoking characteristic biological alterations.

Three main patterns of biodeterioration can be described:

a) intercrystalline growth, in which fungi grow following the weakest parts along marble crystals; this modality of colonisation usually brings to the detachment of crystal (Plate 2b);
b) biopitting in which the fungal colonies grow preferentially in the cavities (Plate 2c); Sterflinger and Krumbein (1997) showed both in natural as well as in laboratory conditions the ability of MCF to produce pits (range 0.3 - 1.5 cm diameter) on marble surface. In natural conditions, different sizes of pits were recognised, while in laboratory experiments, after 10 months of alternate dry and wet cycles, 500 μm diameter holes were formed (Plate 2c).
c) growth in already formed cracks and fissures (Plate 1f). Badly preserved marbles are prone to be easily colonised by fungi that penetrate in the cracks, often producing a deepening of the fissures. When horizontal fissures occur in the subsurface, colonies have a higher chance to expand more than in the surface itself.

Similar biodeteriorative processes can be described for calcareous sandstone and granites localised in the Mediterranean climate. Also in this case, MCF use the weakest points to attack rock surfaces: micas particles, in the case of granite (Sterflinger et al., 1996) or natural or secondary porosity in the case of calcarenite (Urzì and Realini, 1998). Very recently Pinna and Salvadori (1999) showed a typical colonisation of black meristematic fungi on fissures and cracks recently filled with acrylic resins and on carbonate deposits.

However, no evidence was found that fungi may be involved in the stromatolite layers formation both in natural outcrops as well as on monuments. In fact, concerning the possibility to directly relate desert varnish to the presence of MCF, Taylor-George et al. (1983) reported that MCF were found both in the desert varnish as well as in unvarnished rocks and the black colour of these fungi was due to the melanin rather than to deposits of manganese; in addition, there was not clear evidence that the MCF could oxidise Mn. However, MCF could act as a site of accumulation of different clay minerals and interact with bacteria often found in the same place. The presence of bacteria, most of them able to oxidise Mn, could better explain the phenomenon of desert varnish. Bacteria were shown to function better as the site of mineralisation in the patina layers formation (Urzì et al., 1999c).

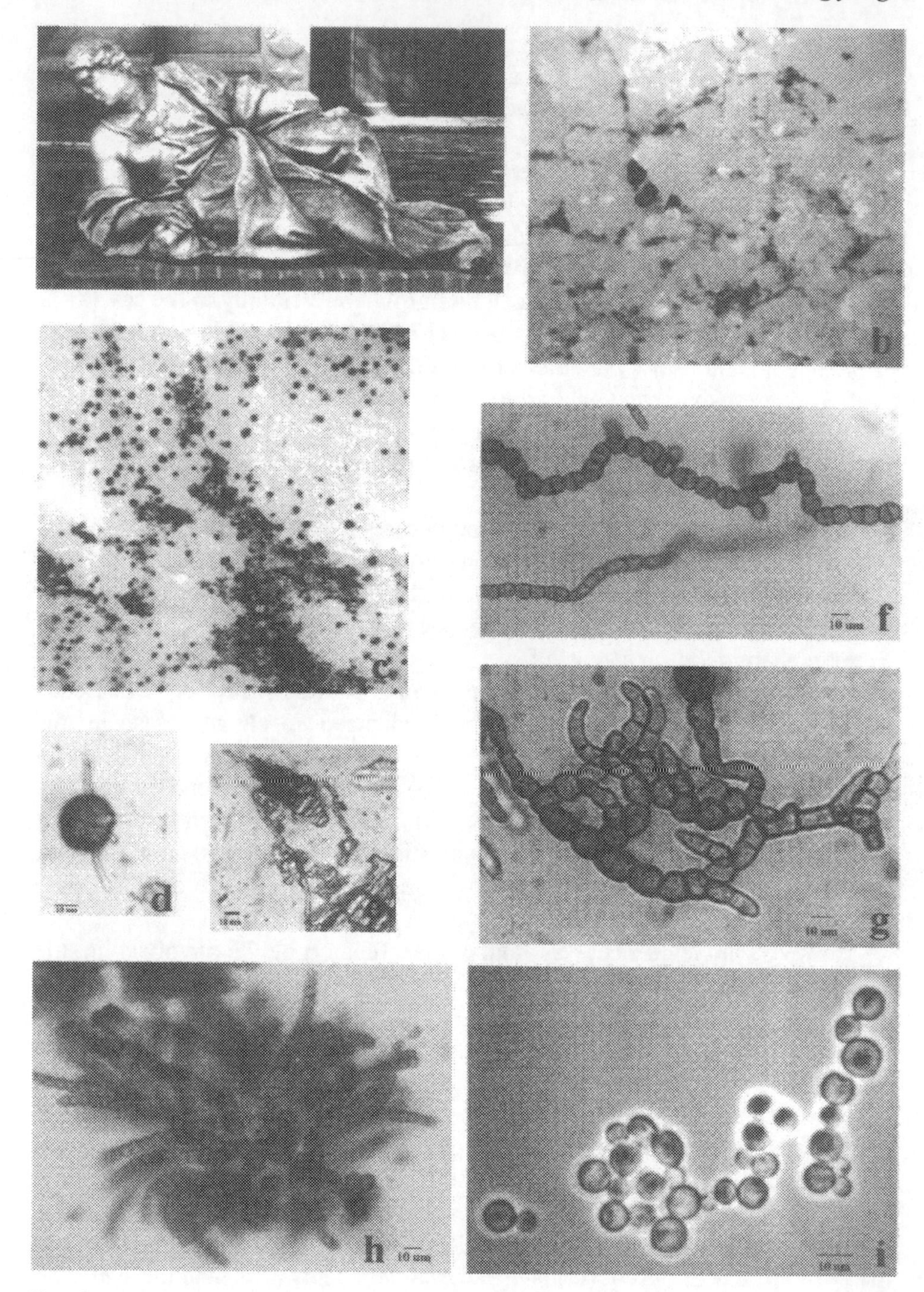

Plate 2

a. Carrara marble statue located in the inner garden of Messina Museum. The whole surface is covered by a black patina due to the fungal growth.

b. Carrara marble surface. Fungal growth between crystals and formation of black clusters. Magnification 80x

c. Delos marble. Diffused biopits all over the marble surface. PAS staining shows extracellular polymeric substances (EPS) also arounds the pits. Mgnification 80x.

d, e. Fungal units producing attachment and/or germinative structures closely attached to the marble surface, evidenced after scotch sampling. Magnification 400x.

f, g, h. Different meristematic black fungi, showing the characteristic morphology. f) *C. apollinis* MC525=CBS100214; g) *C. uncinatum* MC768=CBS100219 and h) *Capnobotryella renispora* CBS572.89. Magnification 400x.

i. Blastic conidia and yeast -like cells of *Sarcinomyces petricola* CBS725.95. Magnification 400x

Since their first isolations, a large number of stone-inhabiting fungal strains were collected and it was recognised that they show some common morphological features that do not allow microbiologists to identify them very easily, especially using classical approaches.

A review of the classical and molecular procedures applied to the taxonomy of black meristematic fungi is presented in the following paragraphs.

2. CURRENT KNOWLEDGE ON TAXONOMY OF STONE-INHABITING FUNGI

Methods applied to fungal taxonomy aim to characterise:

- macro and micromorphology,
- biochemical and physiological properties,
- genotypic characteristics.

Till now black meristematic fungi isolated from rocks have been attributed to the following anamorphic genera: *Exophiala* Carmichael, *Sarcinomyces* Lindner, *Phaeosclera* Sigler, *Lichenothelia* Henssen, *Trimmatostroma* Corda, *Capnobotryella* Sugiyama, *Hortaea* (Horta) Nishimura & Miyaji, *Phaeococcomyces* and *Coniosporium* Link ex Fries. However, most of the above mentioned genera are notoriously hard to classify because of their morphological plasticity (Minter, 1987).

2.1 Determination of morphology

Current methods of identification of fungi are based primarily on microscopic morphology. Microscopic techniques are generally useful when applied to *Hyphomycetes*, but become less rewarding with meristematic fungi, which lack pronounced diagnostic features.

For diagnostic aims, emphasis is placed upon characters of conidiophores, conidia and conidial ontogeny. Without well-defined reproductive structures, the recognition down to the species level is hardly possible. In the meristematic growing fungi (including some black yeasts), morphology is applied with difficulty, due to the fact that many species are highly pleomorphic, having anamorph life cycles with widely divergent types of propagation. Reproduction can occur by liberation of propagules, by disarticulation (sarcinic conidiogenesis) or disruption (endogenous conidiogenesis), blastic conidia or budding cells (Plate 2i).

Some species present meristematic growth as the only type of reproduction, consisting of isodiametrically dividing cells (Plate 1e, Plate 2f, g, h) and endoconidiation (Plate 1e, Plate 2f) that do not allow recognition and delimitation of taxa.

Due to the fact that the first step in the identification is based on the morphological characteristics and on the comparison with strains of already known genera, morphology gives only a presumptive identification at the genus level.

Non-morphological techniques are thus needed.

2.2 Determination of ecophysiology

In addition to morphology, some physiological characteristics, such as nitrogen and carbohydrate assimilation, maximum temperature of growth and proteolytic activity are useful in the identification of fungi. Diagnostic keys based upon biochemical and physiological tests were successfully applied to the identification of black yeasts (de Hoog and Yurlova, 1994; Urzì et al., 1999a). However, physiological identification of meristematic fungi isolated from stone is hampered by their extreme metabolic versatility and poor biomass production. Therefore these techniques are unsuitable for routine diagnostics, although they may help to delimit individual species taxonomically (Wollenzien et al., 1997; De Leo et al., 1999).

Ecophysiological tests have proven to be informative in determining range of growth and survival of strains. It was shown that, in metabolically inactive conditions, these fungi are able to survive extreme conditions, such as temperatures above 100°C, despite the fact that they are mesophilic, being unable to grow above 35°C (Sterflinger and Krumbein, 1995; Sterflinger, 1998) or to resist prolonged UV exposure (De Leo et al., 1998).

2.3 Molecular methods

Following the current taxonomic classification, four lines of relationships within the *Ascomycota* have been suggested for the meristematic fungi (Sterflinger et al., 1997):

The family ***Herpotrichellaceae*** (order **Chaetothyriales**). In this family species of *Exophiala* Charmichael and *Sarcinomyces petricola* Wollenzien de Hoog were included (Wollenzien et al., 1997).

The family ***Dothideaceae*** (order **Dothideales**). In this family some species occasionally isolated from rocks are included as e.g.: *Trimmatostroma abietis*, *Aureobasidium pullulans*, *Hortaea werneckii*.

The family ***Capnodiaceae*** (order **Capnodiales**). *Capnobotryella renispora* isolated from a tile (Titze and de Hoog, 1990) belongs to this family.

The family ***Pleosporaceae*** (order **Pleosporales**). *Botryomyces caespitosus*, closely related to *Alternaria* (this latter frequently found on the stones, Wollenzien et al., 1995), is included in this family.

However, *Coniosporium* strains, recently found to be common inhabitants of stones, do not belong to any of the above mentioned families. Recent papers (Sterflinger et al., 1997; 1999) have demonstrated that they cluster as sister clade of the ***Herpotrichellaceae*** family.

Because the teleomorph connections of the fungi cannot be clarified on the basis of morphology, molecular approaches were applied for the identification and clarification of phylogenetic relationships of meristematic fungi (Sterflinger et al., 1997; Sterflinger et al., 1999; De Hoog et al., 1999).

A wide range of molecular methods have been currently applied in the study of biodiversity of meristematic black fungi and black yeasts.

Depending on the aim of the study, PCR (Polimerase Chain Reaction) based techniques, like RFLP (Restriction Fragment Length Polymorphism) analyses and RAPD (Random Amplified Polymorphic DNA) assays as well as partial or complete DNA sequencing were widely used.

RFLP analyses of amplified Small Subunit ribosomal gene (SSU, 18S rDNA) and Large Subunit ribosomal gene (LSU, 5.8S rDNA and interspacers ITS1, ITS2) were utilized todescribe new species (Wollenzien et al., 1997; Sterflinger et al., 1997; De Leo et al., 1999).

Complete sequencing of SSU (18S rDNA) (Sterflinger et al., 1997, 1999), and LSU genes (5.8S rDNA and interspacers ITS1 and ITS2) (De Hoog et al., 1999) were recently applied to phylogenetic studies and to describe the taxonomic position of meristematic black fungi and black yeasts, which are supposed to belong to the class of *Ascomycetes*.

Morphologically very similar meristematic fungi can be phylogenetically distant. In fact, Sterflinger et al. (1999) found that meristematic black fungi have a close affinity with at least three different orders of *Ascomycetes*: *Chetothyriales*, *Dothideales*, and *Pleosporales*; *C. renispora*, previously supposed to belong to the order *Capnodiales* (Sugiyama and Amano, 1987), and *S. petricola*, previously thought to be related to the family of *Herpotrichiellaceae* (order of *Chaetothyriales*) (Wollenzien et al., 1997), were found to cluster within the order of *Dothideales*.

It is worth noting that the *Coniosporium* species, considered as the most common inhabitants of rock and monuments, form a clearly separate branch next to the family *Herpotrichiellaceae* (order *Chetothyriales*), and they probably represent a new hitherto undefined order of the *Ascomycetes*.

Urzì et al. (1999a) applied the RAPD technique to the strains of *A. pullulans* on the basis of their ecological survey. The authors found the possibility to cluster the isolates from Carrara and Thassos marbles, while environmental strains behaved differently. In addition, it was suggested that specific bands generated by the two primers could be used for the building of molecular probes for *in situ* detection of *Aureobasidium* species from the natural habitat.

3. FUTURE PERSPECTIVES

Besides the problems related to the taxonomy of black meristematic fungi, there is an increasing need for methods that allow their recognition in the environment. Problems specific for working with meristematic fungi are due to limitations in:

- viable cell counts and isolation due to different speeds of growth of fungal units in artificial cultures;
- fungal biodiversity and metabolic activity;
- presence of thick cell wall often melanized, that renders very difficult their detection both using fluorescent dyes and molecular probes.

Sequencing data not only provide a useful basis for phylogeny and taxonomy of the meristematic fungi but are also the basis for new molecular methods applied to the study of fungal biodeterioration.

The basic principle of RNA targeted *in situ* hybridisation (FISH) is that a DNA oligomer that is complementary to a region of the rRNA forms a hybrid with the target region inside morphologically intact cells. The hybrids can be detected by epifluorescence microscopy because the probe is marked with a fluorescent dye (Fig. 2).

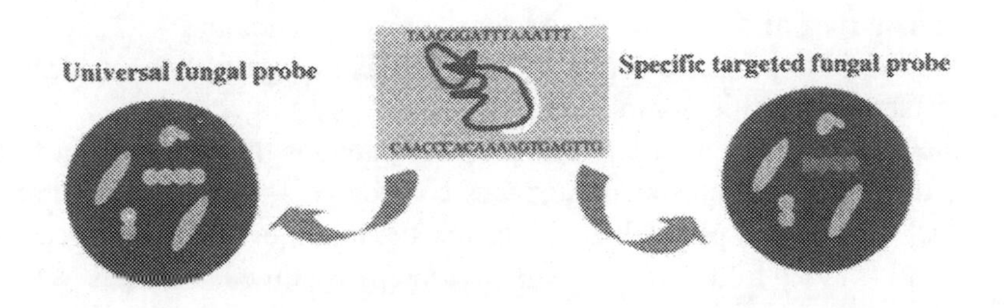

Figure 2. Scheme of FISH hybridisation for fungi.

Probes can be designed for phylogenetically relevant regions of the small ribosomal subunit (SSU) and thus a taxon specific detection of organisms could be possible. Until now the method is mainly used in bacteriology and several protocols and taxon specific probes for bacteria were reported (Amann et al., 1995). Due to the special features of the fungal cell as the rigid cell wall, large vacuoles, the accumulation of secondary metabolites and protein binding sites on the SSU rRNA, the methods cannot be simply transferred from bacteriology to mycology. Very few studies exist on the application of the method to fungi. These studies refer to industrially or medically important yeasts, such as species of *Candida*, *Dekkera* and *Saccharomyces* (Kosse et al., 1997; Lischewski et al., 1996). Li et al. (1997) used the method for the identification of *Aureobasidium pullulans* on leafs. Sterflinger et al. (1998) were the first to present a protocol of *in situ* PCR followed by DNA targeted *in situ* hybridisation in fungal cells. Later on, Sterflinger and Hain (1999) modified the method of *in situ* hybridisation for detecting the black fungi on material surfaces.

At present, some adjustment of the methodologies are still needed for the *in situ* or *ex situ* detection of stone-inhabiting fungi.

4. CONCLUSIONS

This short overview of the taxonomy and ecophysiology of stone-inhabiting fungi shows some of the challenges encountered by scientists working in the field of microbial ecology applied to works of art.

The multifaceted approach to the study of rock-inhabiting fungi has shown the different interesting aspects of this peculiar group of fungi, specialised in colonising rock and monuments surfaces and in particular:

- they seem to prefer to colonise the rock, behaving often as crypto- and chasmo-endolithic organisms;
- they present such a variety of survival strategies that allow them to be often the sole colonisers of surfaces considered as "extreme habitats";
- despite their morphological similarity, the biodiversity of microcolonial fungi is much higher than it was previously assumed.

Molecular studies carried out on the rock-inhabiting black fungi have underlined that they have a polyphyletic origin.

Still much work is needed to create morphologically homogeneous groups coherent with sequencing data. Molecular biology of these groups is hampered by:

1. phylogenetic data are still extremely fragmentary and
2. most reference strains are preserved in herbaria.

Considering the possibility to apply *in situ* and *ex situ* techniques to detect them on monuments surfaces, the broad application of the method is still hampered by the fact that sequencing data in databanks are scarce and thus a lot of work still remains to be done in order to get a good basis for the design of taxon specific probes.

The authors are still working on the topic, so as to better clarify the phylogenetic position of these fungi as well as their role in the deterioration of stones.

ACKNOWLEDGEMENT

This work was carried out with the financial support of the European Community Commission, Contracts N° ENV4-CT98-0704 and N° ENV4-CT98-0707 and of Consiglio Nazionale delle Ricerche (C.N.R.) contract N° 9700720.PF.36.

We like to thank William Fenton, University of Messina, Italy, for his careful revision of the English text.

REFERENCES

Amann, R. I., W. Ludwig and K. H. Schleifer. 1995. Phylogenetic identification and in situ detection of individual microbial cells without cultivation. Microbiol. Rev. **59**: 143-169.

Anagnostidis, K., C. K. Gehrmann, M. Gross, W. E. Krumbein, S. Lisi, A. Pantazidou, C. Urzì and M. Zagari. 1991. Biodeterioration of marbles of the Parthenon and Propylaea,

Acropolis, Athens - Associated organisms, decay and treatment suggestions. *In* D. Decrouez, J. Chamay and F. Zezza (eds.), Proceedings of the 2nd International Symposium. Musée d'art et d'histoire; Genève p. 305-325.

Braams, J. 1992. Ecological studies on the fungal microflora inhabiting historical sandstone monuments. Oldenburg, Ph D. Thesis.

De Leo, F., C. Urzì and G. S. Hoog de. 1999. Two *Coniosporium* species from rock surfaces. Studies in Mycology **43**: 70-79.

De Leo, F., P. Salamone, G. Criseo and C. Urzì. 1998. Crescita di funghi neri meristematici in differenti condizioni ambientali. *In* Atti del Convegno Congiunto SIM-SIMGBM. MonteSilvano Lido, Pescara p. 264.

Eppard, M., W. E. Krumbein, C. Koch, E. Rhiel, J. Staley and E. Stackebrandt. 1996. Morphological, physiological, and molecular characterization of actinomycetes isolated from dry soil, rocks and monument surfaces. Arch. Microbiol. **166**: 12-22.

Garcia-Vallès, M., W. E. Krumbein, C. Urzì and M. Vendrell-Saz. 1997. Colored coatings of monuments surfaces: A result of biomineralization controlled by global climate change or anthopogenic, The case of the Tarragona Cathedral (Catalonia). Appl. Geochem. **12**: 255-266.

Guillitte, O. 1995. Bioreceptivity: a new concept for building ecology studies. The Science of the Total Environment **167**: 215-220.

Hirsch, P., F. E. W. Eckhardt and R. J. Palmer, Jr. 1995. Fungi active in weathering of rock and stone monuments. Can. J. Bot. **73**: 1384-1390.

Hoog de, G. S. and N. A. Yurlova. 1994. Conidiogenesis, nutritional physiology and taxonomy of *Aureobasidium* and *Hormonema*. Antonie van Leeuwenhoek **65**: 41-54.

Hoog de, G. S., P. Zalar, C. Urzì, F. De Leo, N. A. Yurlova and K. Sterflinger. 1999. Relationships of *Dothideaceous* black yeasts and meristematic fungi based on 5.8S and ITS2 rDNA sequence comparison. Studies in Mycology **43**: 31-37.

Kosse, D., H. Seiler, R. Amann, W. Ludwig and S. Scherer. 1997. Identification of yoghurt-spoiling yeasts with 18S rRNA-targeted oligonucleotide probes. System. Appl. Microbiol. **20**: 468-480.

Krumbein, W. E. and C. Urzì. 1993. Biodeterioration processes of monuments as a part of (man-made?) global climate change. *In* M. Thiel, J (ed.), Conservation of stone and other materials. E & FN SPON, London, vol. 1 p. 558-564.

Li, S., R. N. Spear and J. H. Andrews. 1997. Quantitative fluorescence in situ hybridization of *Aureobasidium pullulans* on microscope slides and leaf surfaces. Appl. Environ. Microbiol. **63**: 3261-3267.

Lischewski, A., R.I. Amann, D. Harmsen, H. Merkert, J. Hacker and J. Morschh‰user. 1996. Specific detection of *Candida albicans* and *Candida tropicalis* by fluorescent in situ hybridization with an 18S rRNA-targeted oligonucleotide probe. Microbiol. **142**: 2731-2740.

Minter, D.W. 1987. The significance of conidiogenesis in pleomorphy. *In* J. Sugiyama, (ed.), Pleomorphic fungi: the diversity and its taxonomic implications, Elsevier, Amsterdam p. 8-11.

Pinna, D. and O. Salvadori. 1999. Biological growth on Italian monuments restored with organic or carbonatic compounds. Of Microbes and Art. The Role of Microbial Communities in the Degradation and Protection of Cultural Heritage, Florence, June 1999, International Conference on Microbiology and Consevartion p.149-154.

Staley, J. T., F. Palmer and J. B. Adams. 1982. Microcolonial fungi: common inhabitans on desert rocks? Science **215**:1093-1094.

Sterflinger, K. 1995. Geomicrobiological investigation on the alterations of marble monuments by dematiaceous fungi (Sanctuary of Delos, Cyclades). Oldenburg; Doctorate Thesis.

Sterflinger, K. and W. E. Krumbein. 1995. Multiple stress factors affecting growth of rock-inhabiting black fungi. Bot. Acta **108**: 490-496.

Sterflinger, K., F. Blazquez, M. Garcia-Valles, W. E. Krumbein and M. Vendrell-Saz. 1996. Patina, microstromatolites and black spots as related to biodeterioration processes of granite. *In* European Commission (ed.), Environmental protection and conservation of the European Cultural Heritage p. 391-397.

Sterflinger, K. and W. E. Krumbein. 1997. Dematiaceous fungi as a major agent for biopitting on Mediterranean marbles and limestones. Geomicrobiol. J. **14**: 219 21-230.

Sterflinger, K. 1998. Temperature and NaCl tolerance of rock-inhabiting meristematic fungi. Antonie van Leeuwenhoek **74**: 271-281.

Sterflinger, K. and M. Hain. 1999. In-situ hybridization with rRNA targeted probes as a new tool for the detection of black yeasts and meristematic fungi. Studies in Mycology **43**: 23-30.

Sterflinger, K., G. S. Hoog de and G. Haase 1999. Phylogeny and ecology of meristematic ascomycetes. Studies in Mycology **43**: 5-22.

Sterflinger, K., R. De Baere, G. S. Hoog de, R. De Wachter, W. E. Krumbein and G. Haase. 1997. *Coniosporium perforans* and *C. apollinis*, two new rock-inhabiting isolated from marble in the sanctuary of Delos (Cyclades, Greece). Antonie Van Leeuwenhoek **72**: 349-363.

Sterflinger, K., W. E. Krumbein and A. Schwiertz, 1998. A protocol for PCR *in-situ* hybridization of Hyphomycetes. Internatl. Microbiol. **1**: 217-220.

Sugiyama, J. and N. Amano. 1987. Two *Metacapnodiaceous* sooty moulds from Japan: Their identity and behaviour in pure culture. *In* J. Sugiyama (ed.), Pleomorphic fungi: the diversity and its taxonomic implications. Elsevier, Amsterdam p. 141-156.

Taylor-George, S., F. Palmer, J. T. Staley, D. J. Borns, B. Curtiss and J. B. Adams. 1983. Fungi and bacteria involved in desert varnish formation. Microb. Ecol. **9**: 227-245.

Titze, A. and G. S. Hoog de. 1990. *Capnobotryella renispora* on roof tile. Antonie van Leeuwenhoek **58**: 265-269.

Turian, G. 1975. Maxi-toxitolerance d'une Moissure-Dematiée algicorticole du genre *Coniosporium*. Ber. Schweiz. Bot. Ges. **85**: 204-209.

Urzì, C., W. E. Krumbein and T. Warscheid. 1992a. On the question of biogenic colour changes of mediterranean monuments (coating - crust - microstromatolite - patina - scialbatura - skin- rock varnish). *In* D. Decrouez, J. Chamay and F. Zezza (eds.), Proceedings of the 2nd International Symposium. Musée d'art et d'histoire, Geneve, p. 397-420.

Urzì, C., S. Lisi, G. Criseo and M. Zagari. 1992b. Comparazione di terreni per l'enumerazione e l'isolamento di funghi deteriogeni isolati da materiali naturali. Ann. Microbiol. Enzimol. **42**: 185-193.

Urzì, C. and W. E. Krumbein. 1994. Microbiological impacts on the cultural heritage. *In* W. E. Krumbein, P. Brimblecombe, D. E. Cosgrove and S. Stainforth (eds.), Durability and change: the science, responsability and cost of sustaining cultural heritage, John Wiley & Sons Ltd., London p. 107-135.

Urzì, C. and M. Realini. 1996. Bioreceptivity of rock surfaces and its implications in colour changes and alterations of monuments. Study case of Noto's calcareous sandstone. *In* W. Sand and G. Kreysa (eds.), Proceedings of 10th International Biodeterioration and Biodegradation Symposium. DECHEMA, Monographs vol. 133 p. 151-160.

Urzì, C., S. Trusso and A. Kopecky. 1996. Colonisation patterns of stone surfaces analysed by fractal geometry. *In* J. Reiderer (ed.), Proceedings 8th International Congress on deterioration and conservation of stone; vol. 2 p. 717-723.

Urzì, C. and M. Realini. 1998. Colour changes of Noto's calcareous sandstone as related with its colonization by microorganisms. Internat. Biodet. Biodegrad. **42**: 45-54.

Urzì, C., P. Salamone, F. De Leo and M. Vendrell. 1998. Microbial diversity of Greek quarried marbles associated to specific alteration. *In* M. Monte and R. Snethlage (eds.), Proceedings of the 8th Euromarble EU496 Workshop, Rome, in press.

Urzì C., F. De Leo, C. Lo Passo, and G. Criseo. 1999a. Intra-specific diversity of *A. pullulans* strains isolated from rocks and environment assessed by physiological and molecular (RAPD) methods. J. Microbiol. Meth. **36**: 95-105.

Urzì, C., F. De Leo , P. Salamone and G. Criseo.1999b. Impact of airborne fungi on marble objects exposed at Messina Euromarble site. *In* R. Snethlage (ed.), Proceedings of the 9th Euromarble EU496 Workshop, Bayerisches Landesamt für Denkmalpflege - Zentrallbor Munich p. 39-55.

Urzì, C., M. T. Garcia-Valles, M. Vendrell and A. Pernice. 1999c. Biomineralization processes of the rock surfaces observed in field and in laboratory. Geomicrobiol. J. **16**: 39-54.

Wheeler, M. H. and A. A. Bell. 1988. Melanins and their importance in pathogenic fungi. Current Topics in Medical Mycology. **2**: 338-387

Wollenzien, U., G. S. Hoog de, W. E. Krumbein and C. Urzì. 1995. On the isolation of microcolonial fungi occurring on and in marble and other calcareous rocks. The Science of the Total Environment **167**: 287-294.

Wollenzien, U., G. S. Hoog de, W. E. Krumbein and J. M. J. Uijthof. 1997. *Sarcinomyces petricola*, a new microcolonial fungus from marble in the Mediterranean basin. Antonie van Leeuwenhoek **71**: 281-288.

Yang, C. H. and E. Johanning. 1997. Airborne fungi and mycotoxins. *In* C. J. Hurst, G. R. Knudsen, M. J. McInerney, L. D. Stetzenbach and M. V. Walter (eds.), Manual of Environmental Microbiology. ASM Press, Washington D.C. p. 651-660.

Zezza, F., C. Urzì, T. Moropolou, F. Macrì and M. Zagari. 1995. Indagini microanalitiche e microbiologiche di patine e croste presenti su pietre calcaree e marmi esposti all'aerosol marino e all'inquinamento atmosferico. *In* G. Biscontin and G. Driussi (eds.), Proceedings of XI Convegno Scienze e Beni culturali, La Pulitura dell'Architettura. Libreria Progetto Editoriale, Padova p. 293-303.

MOLECULAR TOOLS APPLIED TO THE STUDY OF DETERIORATED ARTWORKS

Daniele Daffonchio[1], Sara Borin[1], Elisabetta Zanardini[1], Pamela Abbruscato[1], Marco Realini[2], Clara Urzì[3] and Claudia Sorlini[1]
[1]Department of Food and Microbiological Sciences and Technologies, University of Milan, via Celoria 2, I-20133 Milan, Italy; [2]Gino Bozza Center, National Research Council, Piazza L. da Vianci 32, I-20133 Milan, Italy; [3]Institute of Microbiology, University of Messina, Salita Sperone 31, I-98166 Villaggio S. Agata, Messina, Italy.

Key words: biodeterioration, molecular biology techniques, altered artworks

Abstract: Traditional methods normally used to study the microbial populations present on artwork surfaces are time-consuming and often do not reveal the specific characteristics of the microbial ecotypes. The development of new powerful tools from molecular biology (such as polymerase chain reaction [PCR], simple and effective cloning and electrophoretic systems [SSCP, DGGE, TGGE] for nucleic acids separation, automated DNA sequencing facilities and wide databases which rapidly accumulate sequences of signature bacterial genes such as ribosomal genes) have greatly amplified the resolution power of microbial communities analysis (Rölleke et al., 1996; Wollenzien et al., 1997; Moreira and Amils, 1996). In this report, we analyse the contribution of different molecular methods and techniques in investigating the microbial diversity on stoneworks, in relation to the microbe itself and to genes which may play some role in the metabolism of organic air pollutants, and in detecting, by means of molecular probes, particularly dangerous microorganisms whose presence is suspected. A study of the microbial biodiversity of Carrara marble and other stone materials, particularly the presence of *Geodermatophilus*-like organisms, is reported. Among the isolated strains, a strain phenotypically similar to *Geodermatophilus* but genotypically quite divergent from this species has been identified. The partial sequence of the 16S rDNA showed that the strain may represent a new species and even a new genus of the actinomycetes. Moreover, a method to detect directly on the stone samples the presence of genes involved in aromatic hydrocarbon biodegradation is proposed in order to demonstrate the capability of microorganisms to grow by degrading aromatic atmospheric pollutants. Finally, a method is proposed for the detection of *Bacillus* strains.

In particular, a PCR-based system has been developed to monitor directly on stone samples and without isolation the members of the *B. cereus* group, which are widespread in the environment and are frequently found on altered stoneworks (Zanardini et al., 1997). In several cases the bio-molecular techniques proved to be a valid tool for the investigation of artwork biodeterioration.

1. INTRODUCTION

The microbial ecology of altered surfaces is poorly understood even though it may be of great interest and could provide useful information about general and applied microbiology. In fact, analysis of a microbial population can help to understand the steps of colonisation, microbial biodiversity, and the relationship among the different populations on the surfaces and between microorganisms and substrate. It could give data on the total biomass present on the surfaces and identify the most dangerous microorganisms in order to diagnose the pathology, make predictions on the biodeterioration of the substrate, and direct the recovery intervention.

The ecosystems represented by the open air monuments, especially in the Mediterranean area and arid and semiarid regions, may often be considered "extreme" because of exposure to high temperature, sun radiation and drought for long periods of time and to wide temperature variations and high salt concentrations. It is thus possible to encounter, among the microorganisms colonising the artworks, new biovars selected under these conditions and, in some cases, new species as yet unknown. The microorganisms on the surfaces, even those belonging to known species, may have particular physiological and metabolic features not present in strains of the same species, but growing in different environments under other conditions.

In studies on the microbial ecology of artworks, the prime and immediate interest is to evaluate the state of conservation of the artworks and, as a consequence, to address the intervention and the type of restoration. However, from such an investigation another important aspect emerges which goes beyond the sphere of recovery. It concerns the possibility to implement knowledge on microbial biodiversity and to find new microbial strains with metabolic features useful in biotechnology. The new strains could be useful in industrial transformations owing to their thermal resistant proteins, or the capability to be active also at high salt concentrations.

The traditional systems to analyse a microbial community, although still useful for many aspects, are often inadequate. They are time-consuming and often do not reveal the specific characteristics of the most important microbial types. Moreover, cultural methods have proved to be unsuitable to evaluate microbial biomass. In fact, they can detect only 1-10% of the total viable species present in a sample, because most of microorganisms are unculturable (Amann et al., 1995) and cannot be isolated. In the last few years, many investigators have begun to apply molecular biology techniques to study microbial ecology. The development of new powerful tools of molecular biology has greatly amplified the resolutive power of microbial community analysis.

The tools include the following: polymerase chain reaction (PCR); simple and effective cloning systems; high resolving nucleic acids separation procedures such as denaturing gradient gel elctrophoresis (Rӧlleke et al., 1996, 1998), temperature gradient gel electrophoresis (Rölleke et al., 1996, 1998), and single-strand conformation polymorphism (Schwieger and Tebbe, 1998); automated DNA sequencing facilities; wide databases which rapidly accumulate sequences of signature bacterial genes like ribosomal genes. Moreover, the use of suitable genetic markers (e.g. 16S rDNA) has expanded the efficacy of microbial taxonomy and offers excellent tools to type microorganisms and explore microbial diversity. The genes may also be isolated directly from the unculturable microorganisms present in environmental samples, and their sequences allow determination of the phylogenetic group to which the strains belong. Finally, the techniques often allow rapid identification and detection of biodeteriogenic strains that help in the diagnosis of the pathology and in the choice of suitable tools for corrective treatments. In addition to analysis of the microbial ecology of many environments, such as soil, waters, wastewater, food etc., such techniques have also been recently applied to altered artworks in the frame of the study of their conservation.

Molecular tools have been applied by Rölleke et al. (1996, 1998) to the study of microbial communities that are colonising medieval frescoes of the castle of Herberstein (Austria). In this case, the aim was to demonstrate the presence of *Eubacteria* directly in the samples, without isolation of strains in pure culture. PCR was carried out by using a pair of specific primers to amplify *Eubacteria* 16S rDNA. The amplification products underwent DGGE, DNA was re-extracted from the excised bands and re-amplified and the new amplicones were purified and sequenced. With this procedure, members or close relatives of the genera *Halomonas, Clostridium* and *Frankia*, rarely or never described among the microflora colonising works of art, have been identified. The same authors (Rölleke et al. 1996), using the same approach, described a population of *Archaea* in the same wall paintings. In their investigations, the authors demonstrated the presence of *Archaea* in 6 of the 10 samples examined, without isolation in pure culture.

On the basis of the 16S rDNA sequence, the bacteria were assigned to the genus *Halobacterium*. This finding gives an important indication regarding the chemical class of biocides to be used for restoration. In fact, it is known that *Archaea* are not sensitive to some antibiotics effective against *Eubacteria*. In addition to the possibility to discriminate DNA of *Eubacteria* from DNA of *Archaea*, it is possible to choose suitable regions of the 16S rDNA to discriminate microorganisms at a more specific taxonomic level. For instance, Burggraf et al. (1994) designed two 16S rRNA-target oligonucleotide probes for the *Archaea*l kingdoms *Euryarchaeota* and *Crenarchaeota*.

The oligonucleotides fluorescent-probes were utilised also for whole-cell hybridisation, in order to quickly detect the presence of *Archaea* in environmental matrices. Although the method has been applied to pure cultures, it could also be used advantageously for monument materials.

As previously stated, it is possible to isolate from artworks new species of mycetes and bacteria (De Leo et al., 1999; Urzì et al., 1999a). Wollenzien et al. (1997) isolated a new meristematic mycete, *Sarcinomyces petricola*, from a statue in a Messina Museum, from the Dionysos Theatre and from an obelisk at Corfù. On the basis of PCR-ribotyping and physiology, the species has been classified as an anamorph member of the ascomycete family *Herpotrichillaceae* (*Chaetothryales*). All strains belonging to the species have been isolated from open-air marble in the Mediterranean basin.

Sterflinger et al. (1997) isolated from marble in the Sanctuary of Delos two new species of mycetes, named *Coniosporium perforans* and *C. apollinis*. The affinity with other taxa such as *Phaeosclera dematioides* and *Sarcinomyces crustacous* was evaluated on the basis of morphology, 18S rRNA phylogeny and restriction fragment length polymorphism (RFLP) of the internal transcribed spacer region.

Urzì et al. (1999b) demonstrated, on the basis of chemotaxonomic characteristics, that a gram-positive bacterial strain, isolated from Wagenmuller's monument in the old cemetery of Nordfriedhof in Munich (Germany), was not attributable to any known species and, as a consequence, was taxonomically placed between *Nocardioides* and *Aeromicrobium*. The analyses based on the 16S rDNA sequence placed the strain in a new genus. The new species has been named *Marmoricola aurantiacus*.

Another approach to the study of biodeterioration of works of art is the investigation aimed to ascertain the presence of particular biodeteriogenic microorganisms, whose presence is only supposed. For example, it is known that acidogenic bacteria are a serious threat for calcareous stones, particularly if they produce inorganic acids. The study, performed by De Wulf-Durand et al. (1997) on acidophilic bacteria, although not applied to artworks but only to mining environments, could be used as a model for biodeteriorated materials. The authors set up a PCR-based method with primers derived from

16S rRNA sequences to detect the most important microbial groups involved in acidification of the environments: *Acidiphilium, Thiobacillus thioxidans, T. ferrooxidans, T. caldus, Leptospirillum ferrooxidans* and *Sulfobacillus.* The method proved to be very effective and sensitive due to a nested PCR assay which increased the sensitivity.

The investigations carried out by Selenska-Pobell et al. (1998) could be useful for the study of deteriorated materials. The authors focused their attention, among the sulphur-oxidizing bacteria, only on the *Thiobacillus* genus, which is the most common genus of the functional group on the monuments. They applied different molecular biology techniques for the identification and taxonomical categorisation of the genus. The procedure included: i) RFLP analysis of PCR-amplified 16S rDNA and 23S rDNA; ii) intergenic spacer rDNA (between the 16S and the 23s rRNA-genes) amplified ribosomal DNA restriction enzymes analysis (ARDREA); iii) genomic fingerprinting carried out using RAPD and rep-APD (repetitive primer-amplified polymorphic DNA). All these techniques made it possible to detect the strains of *Thiobacillus.* By comparing the methods, the authors concluded that RAPD and rep-APD methods gave the same results in agreement with RFLP of the highly conserved ribosomal RNA (*rrn*) operons. However, RAPD and rep-APD methods were more informative than ARDREA.

Moreira and Amils (1996) focused their attention on *Thiobacillus cuprinus* and set up a method based on amplification carried out by means of specific primers targeting variable regions of the 23S rRNA coding gene and the 16-23S RNA intergenic space region.

Useful suggestions could derive from investigations carried out on pure cultures or environmental matrices, other than the artworks, which may be proficiently adapted to the study of deterioration and conservation of monuments. For example, Hastings et al. (1997) set up molecular biology-based techniques for the direct detection of ammonia-oxidising populations in the soil. They utilised 16S rDNA oligonucleotide probes labelled with digoxigenin in order to hybridise the amplified DNA products obtained with specific primers for *Nitrosomonas* and *Nitrospira.* Moreover, the authors showed that it was possible to detect the presence of specific strains belonging to *Nitrosomonas* (in this case *Nitrosomonas europea*) by utilising a gene probe constructed to detect the presence of the ammonia monooxygenase gene (*amoA*), which codes for the key enzyme of ammonia oxidation.

Schramm et al. (1998), working on bioreactor sludge samples, identified the members of *Nitrosospira* and *Nitrospira* groups by fluorescent in situ hybridization (FISH). To identify the nitrite-oxidizing bacteria, a 16S ribosomal DNA clone library was constructed and screened by DGGE, and selected clones were sequenced. The bacteria represented by these sequences clustered in two groups belonging to *Nitrospira moscoviensis.*

2. A SYNTHESIS OF LABORATORY INVESTIGATIONS BY MOLECULAR BIOLOGY TECHNIQUES

In this paper we analyse the contribution that different molecular methods and techniques may give in investigating the microbial populations on artworks in order to: i) describe the biodiversity as a function of the geographic distribution and the ecotypes of the microbial strains; ii) detect the potentially most biodeteriogenic microorganisms; iii) find genes encoding catabolic enzymes involved in the colonisation and deterioration of artistic materials.

2.1 Biodiversity of *Geodermatophilus*-like strains in relation to the geographic origin and the ecotype

Actinomycetes belonging to the genus *Geodermatophilus* have been isolated from artistic stone works located in the Mediterranean basin and in other areas characterised by a hot climate. The bacteria belonging to the *Geodermatophilus* genus are characterised by a high capacity to adapt to unfavourable conditions. The microorganisms concur in biodeterioration by the production of black or orange pigments, often responsible for chromatic alterations on the surface. Some authors (Krumbein, 1992; Urzì et al., 1992; Eppard et al., 1996) have reported that the pigments produced by *Geodermatophilus* strains may contribute to the development of chromatic alterations such as black or orange patina on the stoneworks. These microorganisms might attack the substrate also by producing organic acids, and they might participate in the formation of crusts on the surfaces by production of extracellular polymeric compounds or by precipitating calcite crystals. *Geodermatophilus* cells show variable morphology (from coccoid to rod form), and they are able to form clusters with strong cell walls. The bacteria are Gram positive, strictly aerobic, chemoorganotrophic, catalase positive and mesophilic. The genus, which is phylogenetically related to the family *Frankiaceae* and especially to the genera *Blastococcus* and *Frankia* (Eppard et al. 1996), includes only one species, *Geodermatophilus obscurus*, divided into several subspecies (such as *obscurus, amargosae, uthahensis*). Little is known about the genus, which has been the object of only a few studies.

In our study we investigated the biodiversity of several strains of *Geodermatophilus* isolated in different regions of the Mediterranean basin. Some of the strains included in the study were isolated from Carrara marble from quarries in Tuscany, Italy, and from test specimens of other stones exposed for 6 years to the polluted air of Milan (Fig. 1).

Other strains were isolated from calcareous sandstone quarries in Noto, Sicily (Italy), and others derived from other stone sampled in Bari (Italy) and Delfi (Greece). All the strains showed many phenotypic characteristics in common with those of the *Geodermatophilus* genus. The strains were analysed by RFLP of the 16S rDNA (ARDRA - amplified ribosomal DNA restriction analysis). The electrophoretic fingerprintings of the strains obtained by ARDRA were compared with those of other strains isolated from different areas and with *Geodermatophilus* strains G20 and G6, two reference strains chosen as a control. On the basis of the bands obtained, it was possible to design a dendrogram in which the strains from Tuscany clustered in one group, whereas the strains from Sicilian stones clustered in another group. It was thus possible to discriminate two groups of strains in relation to their geographical origin (Fig. 1).

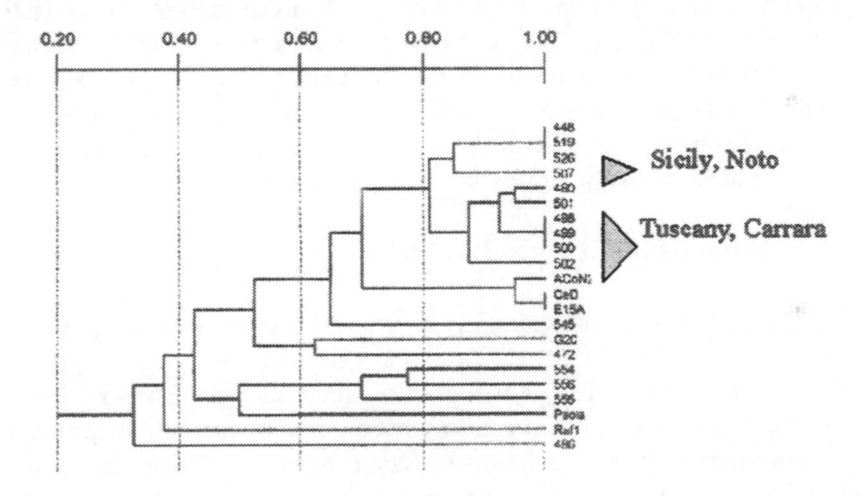

Figure 1. Dendrogram obtained by the statistical elaboration (UPGMA), using the Jaccard coefficient, of the 16S rDNA restriction with AluI, EcoRI, HhaI, RsaI. *Geodermatophilus*-like strains: BC448, BC460, BC498, BC499, BC500, BC 501, BC502, BC507, BC519, BC526, BC545, BC554, BC555, BC496, G20; *Frankia*: AcoN24D, CeD, E15A; *Micromonospora*: 472; *Streptomyces*: Paola; *Rhodococcus*: Raf1. (We thank Marco Bosco and Erica Lumini for providing DNA of frankia strains

In order to add to the knowledge on the genus, three strains, namely the strain Pam1 (isolated in Milan) and the two strains placed at the extremities of the dendrogram, BC448 and BC496, were selected for 16S rDNA sequencing. The partial sequence of the 16S rDNA (about 400 bp) showed that the strains Pam1 and BC448 have a sequence homology of 99% with the reference strain *Geodermatophilus* and of 96% with *Blastococcus aggregatus*, respectively.

Strain 496 showed a very low homology (87.6%) with *Geodermatophilus*, suggesting that the strain may represent a new species or even a new genus among the actinomycetes related to *Geodermatophilus*.

This result shows that strain 496, although phenotypically similar to *Geodermatophilus*, genotypically is quite divergent from the species. The preliminary data indicate that *Geodermatophilus*-like organisms may be easily isolated from stone materials even in the presence of alterations such as in the case of strain Pam1, isolated from an altered Saltrio stone after exposure to the polluted atmosphere of Milan for a prolonged period of time. Moreover, the wide diversity indicated that the molecular variation in the bacterial group is extensive, and collection of the bacteria may be enlarged by more careful isolation attempts. It is noteworthy that isolates from the same geographic location and from the same materials presented similar restriction patterns of the 16S rDNA. This fact could have been due to the selection of particular microbial lineages due to the different environmental conditions. Although they need to be substantiated by a larger number of strains, these results show how molecular biology tools can help to type different lineages of alterative bacteria in a relatively short period of time.

2.2 Detection of target bacteria

An applied aspect of microbial investigations on artworks is the possibility to detect the most biodeteriogenic strains among colonising microorganisms. In fact, it is known that, although all microbial populations have a deteriorating role, some microorganisms have a higher capacity of producing alterations: nitrifying and sulphur-oxidizing bacteria, which are able to produce inorganic acids, filamentous bacteria and mycetes able to penetrate with their hyphae the porous material; spore-forming bacteria able to resist adverse environmental conditions; biofilm-forming bacteria whose exopolymers make them more resistant to biocides.

In our preliminary experiments we focused on the members of *Bacillus cereus* group (Zanardini et al., 1997), which were chosen as model strains from biodeteriogenic microorganisms for three reasons: i) because they are aerobic spore-formers and the spores are characterised by high resistance to environmental stress such as high temperature and pressure, and anti-microbial chemicals; ii) strains belonging to this group are ubiquitous. Indeed, *B. cereus* is known to be a common inhabitant of soil. However, because of its capacity to survive in adverse conditions, it may be encountered in different matrices: water, wastewater (the bacteria may cause gastro-enteritis due to the production of emetic and diarrhoea toxins), soil, airborne microflora, and as contaminants of food products such as rice, milk and milk derivatives (Lechner et al., 1998). Moreover, these bacteria have also been found in the deteriorated materials of some stone monuments (Praderio et al., 1993; Sorlini et al., 1994). iii) finally, because it is a widely studied group (for its implication in food-borne diseases) also under biological molecular aspects. *B. cereus* is strictly related to other spore-forming bacteria, *B. thuringiensis*, *B. anthracis*, *B. mycoides*, *B. wheienstephanensis* (Ash et al., 1991; Lechner et al., 1998) and *B. pseudomycoides* (Nakamura, 1998), which are all included in the so-called "*B. cereus* group". Because of the close genotypic and phenotipic similarities they are not easily distinguishable (Drobniewski, 1993).

Two PCR-based methods have been developed to discriminate *B. cereus* from strains of other species. We applied the first method (*cerA* PCR) to detect *B. cereus* on altered stoneworks; the second method was developed for the specific detection of *B. cereus* as well as for differentiation of the isolates. With the second method it was possible to discriminate strains isolated from different monuments, giving the possibility of correlating particular bacterial haplotypes with the type of alteration.

A) Amplification of the *cerA* locus. The amplification of *cer*A proved to be positive only for the strains of the *B. cereus* group, whereas the other *Bacillus* species were negative in all the PCR experiments, thereby confirming the data obtained by Schraft and Griffith (1995). The limitation of the protocol is related to the ineffectiveness in separating the four aforementioned species.

We analysed sequence polymorphisms among the species of the *B. cereus* group by restriction analysis, but no correlation was found between the restriction haplotypes observed and the taxonomic status of the strains (data not shown). Hence the method may be used as a first screening for monitoring the strains of the *B. cereus* group present in the environment. Further characterisation of the strains and species attribution must be done by other methods.

The effectiveness of the amplification of *cer*A to detect *B. cereus* directly on stoneworks was tested on samples of 15 different altered artworks. The expected bands of 400 bp were obtained in all the samples from which *B. cereus* has been previously isolated (Table 1). In the Cà d'Oro samples, the microorganism was isolated from two of the samples collected from the facade. Amplification was obtained on fresh and on old samples collected five years before and stored in sterile and dry conditions. The PCR results confirmed the high resistance of *B. cereus* spores under stress conditions. Amplifications carried out on the DNA directly extracted from powdered materials were obtained only when *B. cereus* was present in a vegetative stage. No amplification was obtained in the case of spores, probably due to the fact that the spores are very resistant to the treatment for extracting DNA.

The PCR technique allowed the rapid identification and detection of the members of the *B. cereus* group in the presence of extraneous DNA in less than 24 h and using a very small sample of artwork.

The PCR method offers a great advantage as compared to the standard conventional method used to isolate *B. cereus*, since the enrichment culture, followed by isolation and identification, requires at least one week (Zanardini et al., 1997).

Table 1. Detection of *B. cereus* from altered stoneworks by isolation and by PCR amplification of *cer*A.

Sample	Artwork	City	Type of alteration	Isolation of *B. cereus*	Direct PCR	Pre-enrichement and PCR
CE 1	Certosa facade	Pavia	Red spots	+	-	+
CE 2	Certosa facade	Pavia	Red spots	+	-	+
CE 3	Certosa facade	Pavia	Red spots	-	-	-
VE 5	Cà d'Oro facade	Venice	Black deposits	+	-	+
VE 6	Cà d'Oro facade	Venice	Black deposits	+	-	+
VE 10	Cà d'Oro facade	Venice	Black deposits	-	-	-
VE 11	Cà d'Oro facade	Venice	Black deposits	-	-	-
AT 8	Adriano's library	Athens	Chromatic alterat.	-	-	-
AT 9	Adriano's library	Athens	Chromatic alterat.	-	-	-
OR A	St.Brizio's chap.	Orvieto	Degradation	-	-	-
OR I	St.Brizio's chap.	Orvieto	Degradation	-	-	-
LE 3	Roman amphitheatre	Lecce	Chromatic alterat.	-	-	-
MI 1	Senato palace	Milan	Chromatic alterat.	-	-	-
MI 2	Senato palace	Milan	Chromatic alterat.	+	-	+
CA 1	Carrara marble	Carrara	Chromatic alterat.	-	-	-

The isolation of *B. cereus* was performed by traditional microbiological methods on nutrient agar after pasteurisation of the sample. The direct PCR assay was performed on the supernatant obtained after boiling 500 μl of sample in a lysis solution as reported previously (Zanardini et al., 1997). The pre-enrichment and PCR assay was performed as previously reported (Zanardini et al., 1997) by allowing the bacteria present in the sample to pass from a quiescent form (as spores) to the vegetative form, by an overnight (15 h) enrichment at 30°C in nutrient agar. After the pre-enrichment step, the sample was processed as those for the direct PCR assay.

B) Selection of a RAPD fragment specific for the *B. cereus* group to be used for the typing of strains isolated from altered stoneworks. To develop a DNA marker useful to distinguish the strains of the *B. cereus* group from other *Bacillus,* we retrieved specific DNA traits by applying random amplified polymorphic DNA fingerprinting to a collection of strains of the four strains of the *B. cereus* group isolated from different environmental matrices including altered stoneworks (Daffonchio et al., 1999). When analysed by RAPD fingerprinting, the strains of *B. cereus* showed a very high degree of variability. By using the 10-mer primer OPG-8 (5'-TCACGTCCAC-3') in all the strains of the *B. cereus* group, an amplicon of about 850 bp was found. Southern hybridisation experiments confirmed the specificity of the fragment, since hybridisation signals were found only in the strains of the *B. cereus* group. In order to use the specific fragment, named SG-850, as a PCR-specific marker for strains of the *B. cereus* group, it was cloned and sequenced (EMBL accession number AF036105).

From the sequence, two primers (SG-749f 5'-ACTGGCTAATTATGT AATG-3' and SG-749r 5'-ATAATTATCCATTGATTTCG-3') were designed to obtain an amplicon of 749 bp (SG-749), which was amplified only in the members of the *B. cereus* group, suggesting that the fragment could be used for the rapid identification of the strains (Fig. 2A).

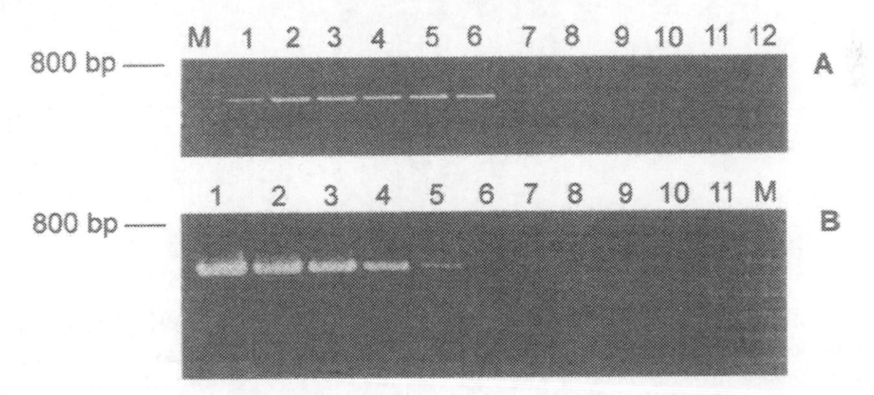

Figure 2. A) Specific amplification of *B. cereus* and relatives using the primers SG-749f and SG-749r. B) Amplification of different amounts of the DNA obtained from *B. anthracis* strain 7700 using the primers SG-749f and SG-749r.

The sensitivity of the PCR reaction, i.e. the lowest quantity of template DNA amplifiable in the selected conditions, was tested by amplifying decimal dilutions of purified DNA from *B. cereus* 31^T and *B. anthracis* 7700. As an example, Fig. 2B shows the amplification sensitivity determined for *B. anthracis* strain 7700. In the condition tested, the lowest quantity of template DNA that could be amplified was 25 and 4.5 pg using the DNA of *B. cereus* 31^T and *B. anthracis* 7700, respectively. The good sensitivity obtained with the PCR test suggests that the method may also be used with small amounts of DNA, like those obtained from environmental samples.

The SG-749 fragments amplified from the DNA of *B. cereus* strains isolated from stoneworks were characterised by restriction analysis (Fig. 3). Different restriction haplotypes were observed among the strains isolated from artworks, indicating that different strains adapted to survive on different stoneworks (Ca' d'Oro in Venice and Certosa in Pavia). The data suggest that the SG-749 fragment constitutes a molecular tool useful to identify *B. cereus* directly from artwork samples without isolation and to type different strains colonising different artworks.

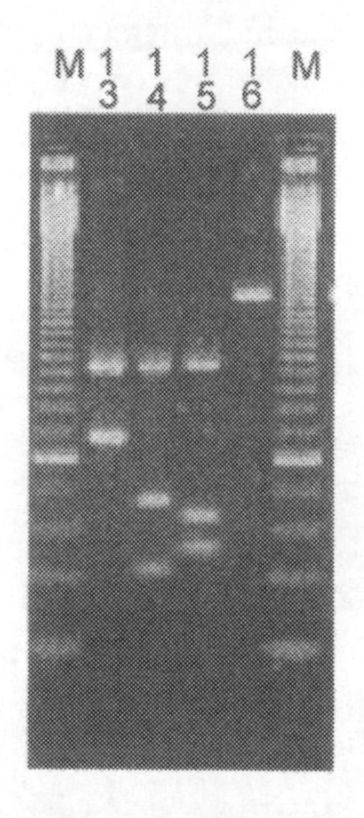

Figure 3. Restriction haplotype found in *B. cereus* strains by cutting the SG-749 fragments with the endonuclease HpaII. Lanes M: 50 bp ladder. Lanes 3 to 6, *B.cereus* strains: DSM 31 T; BC1 isolated from Cà d'Oro; PVBC1 isolated from the Certosa of Pavia; uncutted SG-749 fragment

3. DETECTION OF CATABOLIC GENES INVOLVED IN THE DETERIORATING ACTIVITIES

We elaborated a new conceptual approach to study the causes of biodeterioration. We did not use molecular markers to detect directly biodeteriogenic microorganisms or to identify them. The target of our investigation was the genes which code for particular metabolic activities such as the degradation of air pollutants that may support microbial growth on artwork. The approach could help to understand the mechanisms of the colonisation process and evaluate the impact of colonisers on the colonised surface. In other words, catabolic genes and not whole bacteria were our targets.

To address the investigation, we considered the genes which code for the aromatic hydrocarbon (AHC) degradation pathways. AHC are always found in the polluted atmosphere of industrialised areas and, especially those adsorbed by air particulate, may be deposited on the stone surfaces being in strict physical contact with microbial strains colonising the artworks. The interaction between hydrocarbons in the particulate and microorganisms on the surfaces as concurring causes in stone alteration has been proposed in previous papers (Saiz-Jimenez, 1995; Saiz-Jimenez, 1997).

We focused our attention on the process of colonisation of stones exposed to the polluted atmosphere of Milan. Neither chemolithotrophic nor photolithotrophic bacteria were found on the surface of the stone samples. In contrast, heterotrophic microorganisms were found in a relatively high number, even though the culturable forms showed a relatively low biodiversity. Bacteria were mainly isolated from altered areas of the stone surface, especially from grey and black spots (Fig. 4).

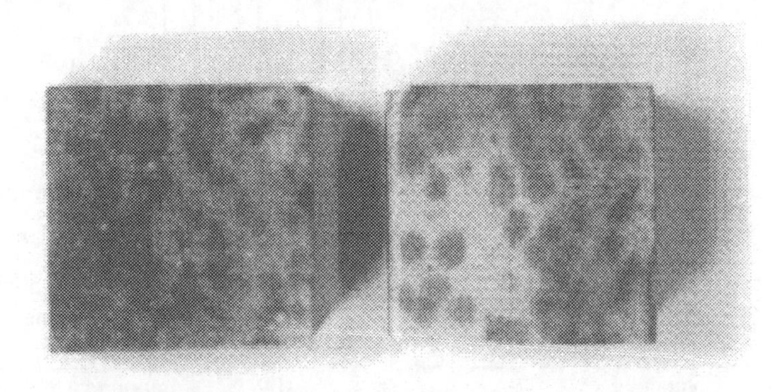

Figure 4. Grey stains on Saltrio stone specimens exposed for 6 years to the polluted air of Milan.

In order to understand whether the microflora was able to grow utilising air pollutant molecules, as previously claimed by other authors (Saiz-Jimenez, 1995; Saiz-Jimenez, 1997), the first step of the investigation was the analysis of the powdered particulate deposited on the surface of the stone sample. The presence of a large number of aliphatic, aromatic and polycyclic compounds was detected (Fig. 5).

In aerobic conditions, AHC may be modified by the oxygenases (mono- and di-), the sole enzymes capable of introducing atmospheric oxygen into the aromatic ring. Among the several classes of the dioxygenases known (dioxygenases play a more important role on the degradation of aromatic compounds than monooxygenases), we chose as the target of our investigation the catechol 2,3-dioxygenase (C23O), a key enzyme in estradiol cleavage of the aromatic ring. In fact, although the aromatic compounds present in the atmosphere are numerous and have different molecular weights (many of them are polyaromatic), the catabolic pathways converge to a small number of compounds, among which catechol is the main and the most common component. The cleavage of catechol and analogous molecules is a bottleneck in the degradation pathways of AHC, and many AHC-degrading bacterial strains so far investigated harbour the enzyme. Many hydrocarbons, including polycyclic aromatics, are degraded via a catechol cleavage step.

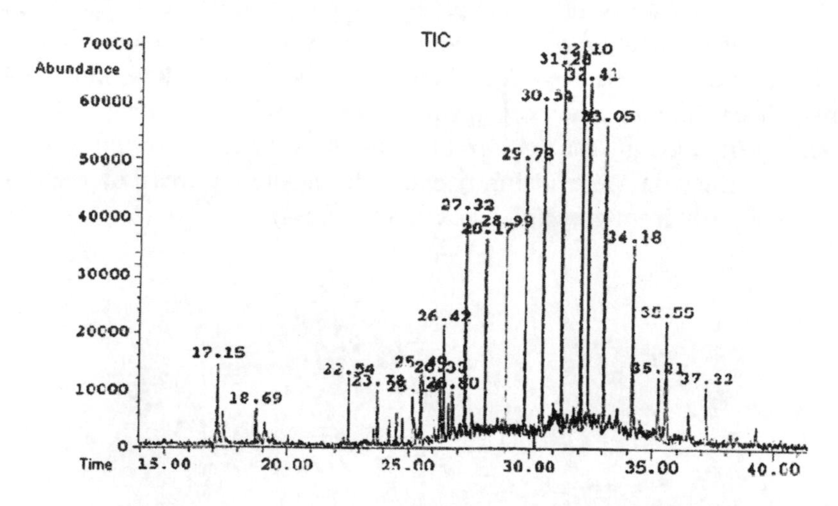

Figure 5. Total ion current (TIC) from GC/MS data for a sample of atmospheric dry deposition. Organic compounds, with relative retention time (RT), are listed below.

RT	Compound
17.15	tridecane
18.69	docosane
20.04	dibenzofuran
22.54	3-methyl 3-cyclohexane
23.78	phenanthrene
25.19	palmitic acid
25.49	pentadecane
26.30	fluoranthene
26.42	octadecanoic acid (me-ester)
26.80	pyrene
27.32	dotriacontane
28.19	octadecanoic acid (et-ester)
28.99	squalene

RT	Compound
29.78	hexadecane
30.54	tetratriacontane
31.28	anthracene
32.10	perylene
32.41	benz[e]acephenanthrylene
33.05	pentatriacontane
34.18	benzo[k]fluoranthene
35.21	benzo[e]pyrene
35.55	3-eicosene
37.22	benz[e]phenanthrene
39.29	hexatriacontane
41.83	eicosane

To study the metabolic potential of microorganisms in relation to AHC degradation, a method to detect the presence of C23O-like genes has been developed. The method is a PCR-based assay that targets the central domain of the C23O genes. The domain is highly conserved between the C23O genes sequences and hence allows the selection of conservative primers useful to amplify all the known C23O genes.

PCR amplification at relatively low stringency was carried out with primers designed from electronic alignment of the nucleotide sequences of known C23O genes. Several amplification products were found, including bands of the expected length (220 bp), using as template DNA extracted from mixed cultures and from pure cultures of five strains able to grow on aromatic compounds and isolated from the black spots of the stone (Fig. 6A). Southern hybridisation experiments confirmed the identity of the bands as C23O-like genes, indicating that microbial strains on stone materials from air-polluted areas harbour the AHC degradation pathways (Fig. 6B).

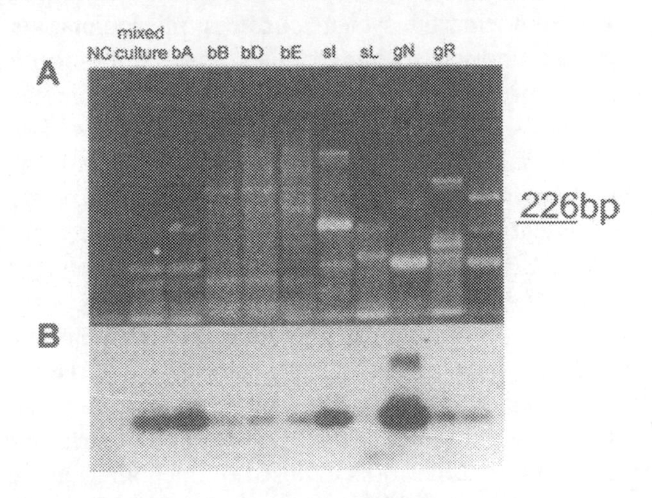

Figure 6. A) Amplification of C23O with primers XylF and XylR of extracted DNA. B) Southern hybridisation. NC negative control, lanes bA, bB, bD, bE, sD, sL, gN, rR represent isolated strains.

The extent of the hybridisation signal varied among the different organisms, indicating that a relative degree of sequence diversity characterises C23O genes present in the microbial population of the altered stones. The presence of C23O-like genes in the microbial cells colonising stones supports the idea that heterotrophic bacteria could be the first colonisers of surfaces. The main problem for heterotrophic bacteria as primary colonisers of stones is the availability of carbon sources for growth. The presence of relatively high amounts of AHC could guarantee carbon source availability only if the bacteria harbour the metabolic pathways for these compounds. The presence of key genes of AHC degradation pathway shows that microorganisms may potentially use the pollutants present on the surface. Such an approach to the study of biodeterioration may supply very useful information on the colonisation and deterioration process by identifying not only microbial groups, but also their metabolic functions.

4. CONCLUSIONS

Molecular biology techniques for the study of the relationships between microorganisms and stoneworks, and artworks in general, are a very useful tool to carry investigations generally known to be very difficult and to acquire new information on the interactions between the artwork surfaces and the microflora. Molecular tools could add to our basic knowledge of the microbial ecology of artworks surface by investigation of the interaction between the different components of a tripartite system: surfaces, microbiota, air.

In addition to the undiscussed advantages for the description of the microbial community present on the artwork surface, such methods can provide very useful tools even from a practical point of view. For example, they can be useful to analyse the recurring microorganisms which may alter artworks in different environments. In fact, such tools are extremely powerful in the diagnosis of potentially alterative microorganisms,since they are rapid, specific and sensitive. Such features make the molecular biology tools very useful also as regards the practical problems that may be encountered in the management of artworks. For example, they may be indispensable to direct the restoration processes when microorganisms partecipate to the development of the alterations. Finally, molecular methods can be used in the detection/selection of new microorganisms of potential industrial interest, due to the ecological features of stoneworks exposed to open air, such as dryness, wide temperature ranges, etc., which stimulate the selection of particular genotypes and phenotypes.

REFERENCES

Amann, R.I., W. Ludwig and K.H. Schleifer. 1995. Phylogenetic identification and in situ detection of individual microbial cells without cultivation. Microbiol. Rev. **59**: 143-169.

Ash, C., J.A.E. Farrow, M. Dorsch, E. Stackebrandt and M.D. Collins. 1991. Comparative analysis of *Bacillus anthracis*, *Bacillus cereus*, and related species on the basis of reverse transcriptase sequencing of 16S rRNA. Int. J. Syst. Bacteriol. **41**: 343-346.

Burggraf S., T. Mayer, R. Amann, S. Schadhauser, C.R. Woese and K.O. Stetter. 1994. Identifying members of the domain *Archaea* with rRNA-targeted oligonucleotide probes. Appl. Environ. Microbiol. **60**: 3112-3119.

Daffonchio, D., S. Borin, G. Frova, R. Gallo, E. Mori, R. Fani and C. Sorlini. 1999. A randomly amplified polymorphic DNA marker specific for the *Bacillus cereus* group is diagnostic for *Bacillus anthracis*. Appl. Environ. Microbiol. **65**: 1298-1303.

De Leo, F., C. Urzì and G.S. De Hoog. 1999. Two new *Coniopsorium* species isolated from rock surfaces. Stud. Mycol. (in press).

De Wulf-Durand P., L.J. Bryant and L.I. Sly. 1997. PCR-mediated detection of acidophilic, bioleaching-associated bacteria. Appl. Environ. Microbiol. **63**: 2944-2948.

Drobniewski, F.A. 1993. *Bacillus cereus* and related species. Clin. Microbiol. Rev. **6**: 324-338.

Eppard M, W.E. Krumbein, C. Koch, E. Rhiel, J.T. Staley and E. Stackebrandt. 1996. Morphological, physiological, and molecular characterization of actinomycetes isolated from dry soil, rocks, and monument surfaces. Arch. Microbiol. **166**: 12-22.

Hastings, R.C., M.T. Ceccherini, N. Miclaus, J.R. Sanders, M. Bazzicalupo and A.J. McCarthy. 1997. Direct molecular biological analysis of ammonia oxidising bacteria populations in cultivated soil plots treated with swine manure. Microbiol. Ecol. **23**: 45-54.

Krumbein, W.E. 1992. L'Acropole - La déteriorration des marbles. Archéologia **280**: 26-31.

Krumbein, W.E., E. Diakumaku and G. Cornelia. 1996. Chemoorganotrophic microorganisms as agents in the destruction of objects of art – A summary. Proceedings of International Congress on Deterioration and Conservation of Stone, Berlin (Germany), 30/9 – 4/10 p. 631-636.

Lechner, S., R. Mayr, K.P. Francis, B.M. Pruss, T. Kaplan, E. Wiessner-Gunkel, G.S. Stewart and S. Scherer. 1998. *Bacillus weihenstephanensis* sp. nov. is a new psychotolerant species of the *Bacillus cereus* group. Int. J. Syst. Bacteriol. **48**: 1373-1382.

Moreira, D. and R. Amils. 1996. PCR-mediated detection of the chemolithotrophic bacterium *Thiobacillus cuprinus* using 23 rDNA- and 16S/23S intergenic spacer region-targeted oligonucleotide primers. FEMS Microbiol. Lett. **142**: 289-293.

Nakamura, L.K. 1998. *Bacillus pseudomycoides* sp. nov. Int. J. Syst. Bacteriol. **48**: 1031-1035.

Praderio, G., A. Schiraldi, C. Sorlini, A. Stassi and E. Zanardini. 1993. Microbiological and calorimetric investigations on degraded marbles from Cà d'Oro facade (Venice). Thermochimica Acta **227**: 205-213.

Rölleke, S., A. Witte, G. Wanner and W. Lubitz. 1998. Medieval wall paintings - a habitat for *Archaea*: identification of *Archaea* by denaturing gradient gel electrophoresis (DGGE) of PCR-amplified gene fragments coding for 16S rRNA in a medieval wall painting. Int. Biodet. Biodegr. **41**: 85-92.

Rölleke, S., G. Muyzer, C. Waver, G. Wanner and W. Lubitz. 1996. Identification of bacteria in a biodegraded wall painting by denaturing gradient gel electrophoresis of PCR-amplified gene fragments coding for 16S rDNA. Appl. Environ. Microbiol. **62**: 2059-2065.

Saiz-Jimenez, C. 1995. Deposition of anthropogenic compounds on monuments and their effect on airborne microorganisms. Aerobiologia **11**: 161-175.

Saiz-Jimenez, C. 1997. Biodeterioration *vs* biodegradation: the role of microorganisms in the removal of the pollutants deposited onto historic buildings. Int. Biodet. Biodegr. **40**: 225-232.

Schraft, H. and M.W. Griffith. 1995. Specific oligonucleotide primers for detection of lecitinase-positive *Bacillus* spp by PCR. Appl. Environ. Microbiol. **61**: 98-102.

Schramm, A., D. De Beer, M. Wagner and R. Amann. 1998. Identification of activities in situ of *Nitrosospira* and *Nitrospira* spp. as dominant populations in a nitrifying fluidized bed reactor. Appl. Environ. Microbiol. **64**: 3480-3485.

Schwieger, F. and C.C. Tebbe. 1998. A new approach to utilize PCR-single-strand-conformation polymorphism for 16S rRNA gene-based microbial community analysis. Appl. Envir. Microbiol. **64**: 4870-4876.

Selenska-Pobell, S., A. Otto and S. Kutschke. 1998. Identification and discrimination of thiobacilli using ARDREA, RAPD and rep-APD. J. Appl. Microbiol. **84**: 1085-1091.

Sorlini, C., O. Salvadori and E. Zanardini. 1994. Microbiological and biochemical investigations on stone of the Cà d'Oro facade in Venezia. 3rd Inter. Symposium on The Conservation of Monuments in the Mediterranean Basin. Venezia, pp. 343-348.

Sterflinger, K., R. De Baere, G.S. de Hoog, R. De Wachter, W.E. Krumbein and G. Hasse. 1997. *Conidiosporum perforans* and *C. apollinis*, two rock-inhabiting fungi isolated from marble in the Sanctuary of Delos (Cyclades, Greece). Antone van Leeuwenhoek **72**: 349-363.

Urzì, C., F. De Leo, C. Lo Passo and G. Criseo. 1999a. Intraspecific diversity of A. *pullulans* strains isolated from rocks and environment assessed by physiological and molecular (RAPD) methods. J. Microbiol. Meth. (in press).

Urzì, C., P. Salamone, P. Schuman and E. Stackrbrandt. 1999b. *Marmoricola aurantiacus* gen. nov., sp. nov., a coccoid member of the family *Nocardioidaceae* isolated from a marble statue. Int. J. Syst. Microbiol. (submitted).

Urzì, C., W.E. Krumbein and T. Warscheid. 1992. On the question of biogenic colour changes of Mediterranean monuments (coating-crust-microstromatolite-patina-scialbatura-skin-rock varnish) *In* D.Decrouez, J. Chamay and F. Zezza (Eds.), The conservation of monuments in the Mediterranean basin, Musée d'Art et d'Histoire, Geneva, pp. 397-420.

Wollenzien, U., G.S. Hoog, W.E. Krumbein and J.M.J. Uijthf. 1997. *Sarcinomyces petricola*, a new microcolonial fungus from marble in the Mediteranean basin. Antone van Leeuwenhoek **71**: 281-288.

Zanardini E., V. Andreoni, S. Borin, F. Cappitelli, D. Daffonchio, P. Talotta, C. Sorlini, G. Ranalli, S. Bruni and F. Cariati. 1997. Lead-resistant microorganisms from red stains of marble of the Certosa of Pavia, Italy, and use of nucleic acid-based techniques for their detection. Int. Biodet. Biodegr. **40**: 171-182.

MOLECULAR APPROACHES FOR THE ASSESSMENT OF MICROBIAL DETERIORATION OF OBJECTS OF ART

Sabine Rölleke, Claudia Gurtner, Guadalupe Pinar and Werner Lubitz
Institute of Microbiology and Genetics, University of Vienna, A-1030 Vienna, Austria

Key words: biodeterioration, DGGE analysis, mural paintings, historical glass

Abstract: One of the most important criteria for the restoration of cultural heritage is the early identification of material deterioration caused by microbial colonization. In order to give guidance to restorers on how and when such restorative efforts are required and to what extent such efforts need to include treatments to stop microbial growth, methods are needed that allow stock taking of microbial communities on the objects of art. It is also important to assess changes in microbial colonization. The present paper describes molecular approaches which allow the monitoring of biodeterioration processes of objects of art, including mural paintings and historical glass.

1. INTRODUCTION

There is ample evidence that microorganisms play an important role in the deterioration of objects of art, particularly if such materials are exposed to open air (Krumbein, 1968; Sorlini et al., 1987; Bock and Sand, 1993). Efforts to eliminate microorganisms which contribute to the deterioration process will be ineffective without a better understanding of microbial diversity. With conventional microbiological means only a small proportion of microorganisms can be cultivated under laboratory conditions (Giovannoni et al., 1990; Ward et al.1990). The cultivation under laboratory conditions of a limited number of microbes may lead to an inaccurate understanding of the species composition and interactions of microorganisms in the biodecay process, resulting in insufficient or even counter-indicated restoration and protection measures (Bianchi et al., 1980).

The methods which are required to analyse the microbial composition will have to allow for the following:

1. A monitoring system that would allow the identification of microorganisms on an ongoing basis. Most systems used currently allow only for an ad-hoc identification of microorganisms, often based on case studies without sustainable impact evaluation.
2. An identification system that will allow for standardized comparison. Currently, conventional methods, such as microbial analysis, vary largely, thus offering different results, influenced by the method that was applied. In addition to sometimes wrong or insufficient results, conventional means do not allow for a better understanding of general features of microbial deterioration
3. A sustainable impact evaluation method that gives restorers an indication of whether their effort to restore an object of art will have long-lasting preservation effects. Such information is important because the treatment of an object of art by, for example, biocides may result in a change but not in the complete elimination of a microbial community. The change in the microbial composition, however, may result in the growth of other microorganisms, which may even be more harmful to the object of art than the microorganisms previously dominating the community. At the same time, it is common knowledge that the use of certain restoration materials may support the growth of microorganisms that in turn will deterioriate the object of art in a medium or long term run.
4. The method should allow for investigation even if only very limited sample amount is available

2. MOLECULAR APPROACHES TO STUDY MICROORGANISMS

A few years ago, DNA based identification techniques of microorganisms have been introduced to investigate the microbial diversity in different environmental habitats (Giovannoni et al., 1990; Ward et al.,1990). The use of such methods demonstrated that conventional microbiological efforts fail to isolate all microorganisms present in natural samples, and that only a small fraction of all microorganisms have been isolated so far. Comparisons of viable and culturable cells in different habitats revealed that the percentage of culturable cells for most environments is below 1% (Amann et al. 1995). There are three different cell types contributing to the non-culturable majority. (1) The bacteria might be known, but the applied cultivation strategies are not suitable for them. (2)

Bacteria might have entered a reversible non-culturable state (3) Most bacteria have never been isolated under laboratory conditions due to a lack of suitable cultivation methods.

Unlike under laboratory conditions, in nature, microorganisms almost never occur in pure cultures. They rather grow in more or less complex microbial communities. The newly developed scientific field of "Molecular Microbial Ecology" focuses on the identity of individual microorganisms and their metabolic potential. New DNA- and RNA-based techniques have been developed, which allow the identification of individual microbial species in sample material without their prior cultivation. The application of molecular biological techniques offers new opportunities for the analysis of individual microorganisms and their composition in sample material.

3. INVESTIGATION OF OBJECTS OF ART USING MOLECULAR APPROACHES

Since the mid-1990s, methods that allow identification of microorganisms without their prior cultivation have been applied to investigate deterioration processes on objects of art (Rölleke et al., 1996, 1998, 1999). The strategy to identify microorganisms in sample material without cultivation comprises: (1) The extraction of genetic material; (2) The amplification, using PCR, of marker DNA sequences; (3) The production of genetic fingerprints; and (4) The identification of microorganisms by comparative sequence analysis.

3.1 Extraction of DNA/RNA from sample material

The extraction of genetic material is extremely important. The method has to be effective for all microorganisms. Cells which cannot be lysed efficiently will not be detected in the subsequent analysis. Therefore, we use an extraction protocol, which combines an enzymatic lysis of cells with a chemical and physical treatment. An additional important aspect is the quality of the extracted DNA. Many of the samples obtained from objects of art contain substances which could inhibit the PCR. To avoid false negative results caused by such inhibitors, we purified extracts from objects of art before using them as template DNA for amplification.

3.2 Amplification by PCR

We amplified selected marker sequences by using the PCR (Polymerase Chain Reaction). This method produces enough sequence material for further analysis from only few target DNA sequences, within a short time. The choice of primers and target sequences is most important when applying molecular approaches. Mostly, ribosomal sequences are used for the identification of microorganisms, for the following reasons: First, they are present in all organisms and contain both variable and highly conserved regions. This allows us to distinguish organisms at all phylogenetic levels. Secondly, an ever growing set of data is entered into databanks, which can be used to compare DNA sequences of known organisms with the ones obtained from the experiments. Ultimately, this allows for the phylogenetic identification.

3.3 Genetic fingerprinting of microbes

As a result of the amplification by PCR, a mixture of DNA sequences, representing the different organisms in the sample material, is obtained. Before individual microbes can be identified, the different marker sequences have to be separated. This can be done by Denaturing/Temperature Gradient Gel Electrophoresis (D/TGGE) of PCR-amplified gene fragments coding for rRNA (Muyzer et al., 1993). This method allows the separation of DNA fragments of identical length but different sequence due to their different melting behaviour in a gel system containing a gradient of denaturants (chemical or temperature). As a result of such an electrophoresis a band pattern is obtained, in which each band represents a microbial taxon in the original sample (see Figure 1). Individual bands can be excised from the gel and sequenced. The potential of this method is such that band patterns already reflect the complexity of microbial communities in the sample. In addition, individual members of the community can be identified. Furthermore, several samples can be run on one gel simultaneously, which allows to compare microbial communities in different samples.

3.4 Identification by comparative sequence analysis

Individual members of microbial communities can be identified by obtaining the sequence information from DGGE bands. These sequences can be analysed by using the EMBL databank (Pearson, 1990) and the RDP-server (Larsen et al., 1993) for comparison with known microbial relatives. While there is a variety of DNA databanks, we used EMBL because it contains almost all sequences, including sequences of unidentified organisms. In addition, the sequences are sent to the RDP server, a specialized databank for ribosomal sequences. A reliable phylogenetic identification should be based on sequence comparison using at least two databanks, as they may contain different sequences.

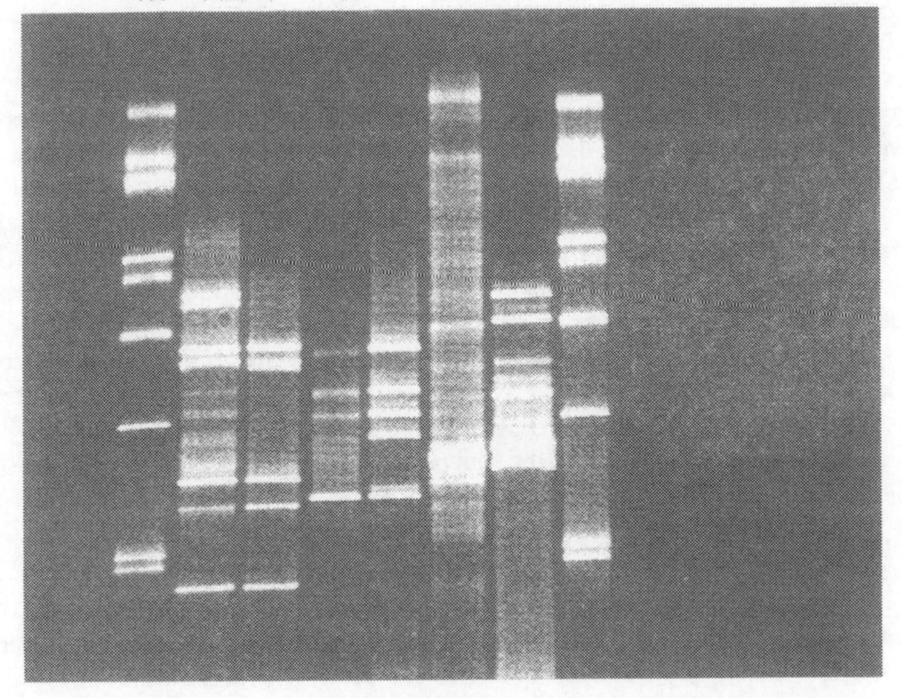

Figure 1. Ethidium bromide stained DGGE separation pattern of DNA fragments coding for 16S rRNA. These fragments were amplified from the Herberstein wall painting DNA, by using a bacterial specific primer pair (HB 4, HB 1 and HB7). Lane M represents a marker containing the ribosomal fragment of known bacteria.

4. ANALYSIS OF OBJECTS OF ART

4.1 Mural paintings

Like other cultural heritage, also mural paintings can support the growth of microorganisms, that contribute to the biodeterioration of the paintings and their grounding. Bacteria which, *inter alia*, grow on the surface of such paintings might discolor the painting not only through their own pigments, but also by excreting metabolic products. Heterotrophic bacteria can use organic compounds as growth substrates, producing acids which cause discoloration of the paint, or change its consistency. In addition, mycelia of fungi and actinomycetes can penetrate into the painting and its grounding resulting in the mechanical destruction of the object. Current restoration efforts can conceivably have an opposite effect, particularly if these treatments use substances which support the growth of microorganisms and, consequently, accelerate the deterioration process. Often not sufficiently considered, microbial colonization should be taken into consideration when planning the restoration of ancient wall paintings. Therefore, an inventory of the existing microorganisms associated with the damage of the paintings is a prerequisite for including biodecay as an integral part of the restoration process. Most bacteria which cause damage to wall paintings cannot be cultivated under laboratory conditions. To trace the non-cultivated fraction of bacteria, molecular biological means were applied to identify microorganisms in a biodegraded wall painting from the 13th century (Rölleke et al., 1996 and 1998). This approach demonstated the presence of a variety of microorganisms, which had never been isolated from wall paintings. Especially halotolerant bacteria, grouping with members and close relatives of the genera *Halomonas* and *Halobacterium* were identified. Extremely halophilic microorganisms had never been identified in these environments. Figure 1 shows a DGGE-profile of samples taken from locations of different biodeteriorated areas of the wall painting.

4.2 Historical window glass

The first report of biodeterioration of glass dates back to the beginning of this century (Mellor 1924). However, until now only a few studies have been published concerning the microflora and the corrosion mechanisms of glassy surfaces. It can be assumed that microorganisms growing on glass must be highly specialized so as to allow for sufficient nutrition for their metabolism. Since the exact conditions cannot be imitated in the laboratory, the application of conventional techniques may lead to the investigation of microorganisms, which do not fully reflect the actual bacterial population of the window glass. Conservation treatments as well as environmental conditions and glass composition are considered to be additional causes of microbial attack (Drewello, 1998). Leaching, the formation of etched traces, pitting, mineralisation, and crust formation are caused by the mobilisation of mono- and divalent metals of the glass and are well-documented problems associated with the growth of microbes on glass surfaces. The application of molecular methods to analyse microbial communities on historical glass demonstrated that a variety of bacteria is associated with deterioration processes of historical glass (Rölleke et al., 1999) and that only little is known regarding the composition of the microbial flora that may play a role in the deterioration process of glass. The investigation of bacterial communities on a 19th century soda-lime silicate glass panel from the German protestant chapel in Stockkämpen revealed complex band patterns (data not shown). This complexity suggests that a large variety of bacteria exists in the samples as part of a complex microbial community. To identify the bacteria present in the individual bands, bands were excised from the gel and sequenced. The results of the comparative sequence analysis are shown in Table 1. One of the most dominant bands was found to represent members of the order

Cytophagales and the highest similarity in partial 16S rDNA was obtained with species of the genus *Flexibacter.* Bacteria of this group grow aerobically and are microaerophilics or facultative anaerobe. They are all organotrophs. Many of these bacteria are able to degrade biomolecules. The *Cytophagales* are found in a wide spectrum of habitats, mainly those rich in organic material. However, they can adapt also to low nutrient levels. It is likely that, when growing on historical glass, these bacteria live on a biofilm, which is produced by other microorganisms. We could also identify a sequence which shows the highest 16S rDNA similarity with sequences of the genus *Nitrosospira* (Table 1). To our knowledge, nitrifying bacteria have never been identified on glass. They could play an important role in biodeterioration processes of glass surfaces, due to their ability to produce inorganic acids. Furthermore, we found several sequences which showed the highest similarity with 16S rDNA sequences belonging to different genera of actinomycetes (see Table 1). Like fungi, many actinomycetes are capable of producing a stable mycelium. The penetration by hyphae of different materials is associated with the ability to excrete a wide range of products, including a variety of enzymes (Weirich, 1989; Williams, 1985). Besides mechanical destruction of the grounding, actinomycetes that produce mycelium can cause different deterioration phenomena, due to their growth and production of metabolites, like acids and pigments (Urzì et al., 1991). Actinomycetes have already been isolated from historical glass surfaces in the past. Biopitted glass samples from the Cologne Cathedral were found to be colonized by fungi and a coccoid bacterium that was identified as *Micrococcus* sp. (Krumbein et al., 1991). Besides a sequence showing 91% 16S rDNA similarity with sequences from the genera *Streptomyces*, *Micrococcus* and *Arthrobacter*, also sequences could be identified representing bacteria belonging to the family *Frankiaceae*. The sequences obtained in this study showed up to 99.1% similarity to sequences of the genera *Frankia* and *Geodermatophylus*. It has been believed that bacteria belonging to *Frankiaceae* occur worldwide, but that they are soil organisms only (Baker and O'Keefe, 1984; Lechevalier, 1989). In 1996, two reports were published which showed that *Frankiaceae* can also be present in other habitats. In a study of the biodiversity of rockdwelling bacteria, *Geodermatophilus sp.* could be isolated from rocks and monument surfaces (Eppard et al., 1996). Also the application of molecular tools for the analysis of a biodeteriorated medieval wall painting in Austria, as described earlier on, demonstrated the presence of members or close relatives of the genus *Frankia* (Rölleke et al., 1996). The identification of members or close relatives of the genera *Frankia* and *Geodermatohilus* on historical glass proves that members of the family *Frankiaceae* can be present also in other habitats than soil.

Table 1. % Similarity of DGGE-bands obtained from historical glass in partial 16S rDNA with sequences of known bacterial relatives in databanks.

Bacterial strain	%
Flexibacter sp.	89.6
Nitrosospira sp	94.3
Arthrobacter sp.	91.0
Streptomyces sp.	91.0
Micrococcus sp.	91.0
Frankia sp.	99.1
Geodermatophilus sp.	96.4
Frankia sp.	97.3
Geodermatophilus sp.	97.3

5. CONCLUSION AND PERSPECTIVES

Our results suggest that, in the past, the use of selective cultivation strategies favoured the identification of those microorganisms, which were easy to isolate under laboratory conditions, but which were not necessarily those causing biodeterioration.

It becomes more and more evident that biodeterioration has to be understood as a result of the activity of an entire microbial community, rather than that of individual members of such a community. Until and unless the composition of such communities is fully known, countermeasures will remain piecemeal efforts, which may even have negative effects on the preservation of cultural heritage.

ACKNOWLEDGMENTS

The investigations described in this communication have been financed by the EU (ENV4-CT98-0705) and by the Deutsche Bundesstiftung Umwelt Osnabrück (AZ:11472).

REFERENCES

Amann, R. I., Ludwig W. and Schleifer, K.-H. 1995. Phylogenetic identification and in situ detection of individual microbioal cells without cultivation. Microbiol. Rev. **59**: 143-169.

Baker, D. and O'Keefe, D. 1984. A modified sucrose fractionation procedure for the isolation of *Frankiae* from actinorhizal root nodules and soil samples. Plant Soil **78**: 23-28.

Bianchi, A., Favali, M. A., Barbieri, N. and Bassi, M. 1980. The use of fungicides on moldcovered frescoes in San Eusebio in Pavia. Int. Biodeter. Bull. **16**: 45-51.

Bock, E. and Sand, W. 1993. The microbiology of masonry biodeterioration. J. Appl. Bacteriol. **74**: 503-514.

Drewello, R. 1998. Mikrobiell induzierte Korrosion von Silikatglas -unter besonderer Berücksichtigung von Alkali-Erdalkali-Silikatgläsern. Thesis, Universität Erlangen-Nürnberg.

Eppard, M., Krumbein, W. E., Koch, C., Rhiel, E., Stanley, J. T. and Stackebrandt, E. 1996. Morphological, physiological, and molecular characterization of actinomycetes isolated from dry soil, rocks, and monument surfaces. Arch. Microbiol. **166**: 12-22.

Giovannoni, S. J., Britschgi, T. B., Moyer, V. and Field, K. G. 1990. Genetic diversity in Sargasso Sea bacterioplankton. Nature **345**: 60-63.

Krumbein, W. E. 1968. Zur Frage der biologischen Verwitterung: Einfluß der Mikroflora auf die Bausteinverwitterung und ihre Abhängigkeit von edaphischen Faktoren. Z. Allg. Mikrobiol. **8**: 107-117.

Krumbein, W. E., Urzi, C. and Gehrmann, C. 1991. Biocorrosion and biodeterioration of antique and medieval glass. Geomicrobiol. J. **9**: 139-165.

Larsen, N., Olsen, G. J., Maidak, B. L., McCaughey, M. J., Overbeek, R., Macke, T. J., Marsh T. L. and Woese. C. R. 1993. The ribosomal database project. Nucleic Acid Res. **21**: 3021-3023.

Lechevalier, M. P. 1989. Actinomycetes with multilocular sporangia. *In* BMSB, S. T. William, M. E. Sharpe and J. G. Holt (Eds.), Vol 4. Williams and Wilkins, Baltimore p. 2405-2417.

Mellor, E. 1924. The decay of window glass from the point of view of the lichenous growth. J. Soc. Glass Technol. **8**: 182-186.

Muyzer, G., de Waal, E. C. and Uitterlinden, A. G. 1993. Profiling of complex microbial populations by denaturing gradient gel electrophoresis analysis of polymerase chain reaction-amplified genes coding for 16S rRNA. Appl. Environ. Microbiol. **59**: 695-700.

Pearson, W. R. 1990. Rapid and sensitive sequence comparison with FAST and FASTA. Methods Enzymol. **183**: 63-98.

Rölleke, S., Muyzer, G., Wawer, C., Wanner, G. and Lubitz, W. 1996. Identification of bacteria in a biodegraded wall painting by denaturing gradient gel electrophoresis of PCR-amplified gene fragments coding for 16S rRNA. Appl. Environ. Microbiol. **62**: 2059-2065.

Rölleke S., Witte A., Wanner G. and Lubitz, W. 1998. Medieval wall paintings: A habitat for *Archaea* - Identification of *Archaea* by denaturing gradient gel electrophoresis (DGGE) of PCR- amplified gene fragments coding for 16S rRNA in a medieval wall painting. Int. Biodeter. Biodegr. **41/1**: 85-92.

Rölleke, S., Gurtner, C., Drewello, U., Drewello, R., Lubitz, W. and R. Weissmann, R. 1999. Analysis of bacterial communities on historical glass by denaturing gradient gel electrophoresis of PCR-amplified gene fragments coding for 16S rRNA. J. Microbiol. Methods **36**: 107-114.

Sorlini, C., Sacchi, M. and Ferrari, A. 1987. Microbiological deterioration of Gambara's frescos exposed to open air in Brescia, Italy. Internat. Biodet. **23**: 167-179.

Urzi, C., Lisi, S., Criseo, G. and Pernice, A. 1991. Adhesion to and degradation of marble by a *Micrococcus* strain isolated from it. Geomicrobiol. J. **9**: 81-90.

Ward, D. M., Welle, R. and Bateson, M. M. 1990. 16S rRNA sequences reveal numerous uncultured microorganisms in a natural community. Nature **345**: 63-65.

Weirich G. 1989. Investigations of the microflora of mural paintings. Mater. Organismen **24**: 139-159.

Williams, S. T. 1985. Streptomycetes in biodeterioration - their relevance, detection and identification. Int. Biodeter. **21**: 201-209.

COMPARATIVE STUDIES OF MICROBIAL COMMUNITIES ON STONE MONUMENTS IN TEMPERATE AND SEMI-ARID CLIMATES

Eric May[1], Sophia Papida[1], Hesham Abdulla[2], Sally Tayler[1] and Ahmed Dewedar[2]

[1]University of Portsmouth, School of Biological Sciences, King Henry Building, King Henry I Street, PO1 2DY, Portsmouth, UK; [2]Department of Botany, Faculty of Science, Suez Canal University, Ismailia, Egypt.

Key words: climatic factors, bacteria, biodeterioration, stone

Abstract: Climate is recognised to play an important part in influencing the activity of microorganisms on stone in monuments and other objects of cultural value. In the UK and Greece, the numbers and distribution of heterotrophic bacteria were not strongly related to seasonal changes in temperature and rainfall. At Portchester Castle, qualitative changes in bacterial populations have been observed; actinomycetes were found only on decayed stone and dominant on stone from Tell Basta, in semi-arid Egypt. In the Minoan Palace at Petrás and the fortifications of Khaniá, higher counts of halotolerant heterotrophic bacteria were found in sheltered areas on stone showing other biological growths and salt efflorescence. For all monuments, stones of lower mechanical strength supported higher bacterial counts and electron microscopy showed extensive sheets of biofilm. Estimations of carbohydrate in stone could also be related to whether the sites were exposed or protected. There is evidence to suggest that variations in the nature of bacterial populations may be dependent on season in temperate regions and in the Mediterranean climate related to the location within the monument. Biofilm production and perhaps halotolerance provide means by which bacteria resist adverse changes in moisture levels. Extreme fluctuations in moisture may induce major shifts in bacterial populations selecting filamentous, spore-forming types that penetrate deeper into the stone.

1. INTRODUCTION

Climate is recognised to play an important part in influencing the activity of microorganisms on stone in monuments and other objects of cultural value. Biodeterioration is a process involving several types of microorganisms, notably bacteria, fungi and algae in combination with lichens or mosses. Early workers suggested that the very presence of living organisms on stone increased its susceptibility to damage through their water-binding capacity. It has recently been suggested that the mineralogy, porosity, surface roughness and capacity to collect water and organic materials will control its bioreceptivity (Guillitte, 1995) and tendency to biodeterioration (Krumbein and Gobushina, 1995). However, environmental factors such as temperature, light intensity, pH and relative humidity will affect the number and types of colonising species (Tiano et al., 1995) and hence the progress of the colonisation.

Water is very important in causing the mechanical degradation of stone, controlling mainly porosity and bulk density (Turk and Dearman, 1986), while its circulation through stone is controlled by its porosity size, distribution, specific surface and capacity (Appolonia et al., 1996). It promotes salt movement resulting in crystallisation and expansion due to changes in relative humidity (Puehringer, 1996) causing high shear strength forces (Fookes et al., 1988), inter-granular disintegration, scaling, flaking and general structural damage (Warke and Smith, 1994). Rainfall intensity (Guidobaldi and Mecchi, 1985) and amount (Smith et al., 1995) greatly influence erosion rates of marbles and limestones while its effect on sandstone is contradictory (Lammel and Metzig, 1997).

The presence and amount of water control biological activity and growth (Krumbein et al., 1996; Silva et al., 1997). Hydrodynamic forces bring bacteria close to substrates and their sorption onto solid surfaces often assisted by the formation of extra-cellular polymeric substances (EPS), or biofilms, whose main role is to adapt them to the harsh environment (Decho, 1994). The protective biofilms secreted by bacteria form a hygroscopic gel which takes up water from air and releases it when the RH is low, thus increasing the moisture content of stone and reducing its porosity (Sand, 1997). When subject to starvation on rocks, they differentiate their size, shape and EPS production to allow deeper penetration and spread (Lappin-Scott and Costerton, 1990). Their penetration time and growth on stone may depend on stone pore size and permeability (Jenneman et al., 1985).

Salts can also affect the property of EPS causing a reduction in viscosity of the charged polymers by shielding fixed charges against mutual electrostatic repulsion and convert an extended stiff form into a smaller and more flexible structure (Christensen and Characklis, 1990). The lower thermal conductivity of biofilms compared to rocks (Bland and Rolls, 1998) combined with their impedance of liquid movement through stone (Sand, 1997) may result in uneven heat transfer during summer or freeze/thaw behaviour in winter (Palmer and Hirsch, 1991).

Thus the responses of microbial populations on stone monuments to variations in climate can result in mechanical damage to stone. The main mechanisms involve biofilm formation in addition to salt stress and contraction/expansion as a result of temperature change. Differences in expansion/contraction coefficients between stone and biofilms may lead to mechanical damage during wetting and drying cycles. This paper compares the bacterial populations that occur on stone monuments in temperate (UK), Mediterranean (Crete, Greece) and semi-arid (Egypt) climates in order to determine how climate may influence the characteristics of the microbial community.

2. EXPERIMENTAL RESULTS

2.1 Study sites

Portchester Castle lies to the north edge of Portsmouth Harbour and the surrounding geology consists of chalk overlaid and penetrated by thin and discontinuous marl. The climate is temperate with moderate monthly rainfall (37.3 – 94 mm) during the year and a mean monthly temperature that varies from 5.9 to 18.2°C during the year and rarely exceeds 20°C in summer (data supplied by the UK Meterological office). The temperature usually reaches freezing point during the winter months.

Petrás Minoan Palatial Building is situated at Siteia, East Crete, Greece. It is situated on a hill by the sea on the edge of a plain. The climate is mediterranean (mean monthly temperatures range from 11.5°C to 25.8°C) with hot dry summers and relatively moist moderate winters (mean monthly rainfall 0 – 80.8 mm). Humidity can be high on windless days and evaporation is high in the summer. Frost and snow occur very rarely (data supplied by the Agricultural Service of Heraclion, Crete, Greece).

Khaniá in West Crete, Greece has a fortified city wall and the Sabbionera bulwark was chosen because of its characteristic position offering protection from north-west winds. The climate is Mediterranean but moister than Siteia (mean monthly rainfall 0.3-113.3 mm). Mean monthly temperatures range from 11.2°C (winter) to 26.2°C (summer). During the summer evaporation is high and on windless days humidity can also be high. In winter, frost occurs about once every four years and snow once every ten years (data supplied by the Agricultural Service of Heraclion, Crete, Greece).

The Tell Basta region near Zagazig City in Egypt has a series of ancient stone tombs and these were chosen for extensive study of the microbiology of the building stone. The climate is semi-arid with no rainfall in the summer and a total mean monthly rainfall of 6.7 mm during the winter months. The mean monthly temperature during the winter is usually 19.4°C rising to 31.1° C during the summer. Relative humidity is low throughout the year ranging from 59.3 (summer) to 62.3 (winter).

2.2 Climate versus counts

A complete analysis of the relationship between climatic factors such as rainfall and temperature and bacterial counts found on stone at Portchester Castle was carried out. Direct counts (obtained using a Helber counting chamber with phase contrast microscopy) for sound stone (Table 1) showed a low positive correlation with temperature and a relatively high negative correlation with rainfall. For decayed stone a low negative correlation was found between direct counts and temperature (r = -0.487) but no correlation was found between direct counts and rainfall (r = -0.056). Counts were of the order of $10^8 – 10^9$ cells per gram stone and sound stone supported a higher population. These counts included viable and non-viable cells, confirmed by activity measurements using INT (Tayler and May, 1995), and direct counts were always much more than plate counts for viable bacteria (usually $<10^6$).

Table 1. Correlation matrix for direct counts and climatic factors

	Sound	Decayed	Rainfall
Decayed	0.029		
Rainfall	-0.752	0.056	
Temperature	0.244	-0.487	-0.310

Viable counts from a range of medium strengths for sound and decayed stone showed very low correlation with rainfall but a higher correlation with temperature (Table 2). There is likely to be considerable variations on the surface exposed stone in a temperate climate. No consistent relationship has emerged between counts of bacteria (total or viable) and either temperature or rainfall. However, Tayler and May (1991) reported that higher numbers of bacteria were isolated during the winter and early spring than in summer and early autumn with no bacteria being recovered in some months. However, when colony appearance, morphology and biochemical tests for the stone isolates were analysed by cluster analysis, particular bacterial types could not be related to seasonal change or the type of stone. In addition, a constant problem in this type of work was the existence of a significant number of bacteria which grew on initial isolation but could not be sub-cultured. This is consistent with the difference between direct and viable counts and a similar phenomenon was observed by Mallory et al. (1977) for isolation of oligotrophs from seawater.

Table 2. Correlation matrix for viable counts* and climatic factors

	Sound	Decayed	Rainfall
Decayed	0.770		
Rainfall	-0.141	0.106	
Temperature	0.651	-0.717	-0.308

* on a reduced-strength (0.1%) acetate medium

The survival of bacteria is dependent on upon a number of factors and in particular temperature and moisture levels can be critical. The importance of the rock surface nanoclimate as a factor controlling microbial colonisation has been reviewed by Friedmann and Ocampa-Friedmann (1984). In the antarctic cold desert the rapid temperature oscillations at around 0°C result in alternate freezing and thawing of the rock surface which inhibits microbial growth. Surfaces receiving direct insolation during the summer months are likely to heat up considerably and also cause drying out of the surface which will severely limit microbial growth. At Portchester this resulted in the emergence of spore-forming bacteria such as *Bacillus* and filamentous actinomycetes as the dominant survivors (Tayler and May, 1991). In a more extensive study of stone monuments in Southern England, to be published in detail elsewhere, surveys confirmed that these Gram-positive organisms are dominant on stone in the summer months.

2.3 Filamentous Bacteria

At Portchester Castle, filamentous bacteria were never isolated from sound stone throughout a study conducted by Tayler (1991) and the number of actinomycete colony types isolated from decayed stone was higher during the winter and spring months than in the summer and autumn. Fewer different actinomycete colony forms were isolated during the summer months. This may be attributed to prolonged drying out of the substrate which has been observed to reduce counts in soils (Jager and Bruins, 1974). The majority of isolates which were isolated only from decayed stone on a starch-casein medium belonged to the genus *Streptomyces* but *Micromonospora* and *Micropolyspora* were occasionally found as well. These findings support an early study by Webley et al. (1963) who recorded that filamentous bacteria were found in relatively high numbers on weathered stones and that they belonged to the genus *Streptomyces.* Other workers (Eckhardt, 1985; Warscheid, 1990) have also reported their presence on stone. Starch-casein medium (Kuster and Williams, 1964) gave a greater range of colony types than arginine-glycerol salts medium (El-Nakeeb and Lechevalier, 1963).

Actinomycetes have also been isolated from stone samples taken from a tomb at Tell Basta, Zagazig City, Egypt where the rainfall is very low throughout the year and seasonal temperatures are high. The data in Table 3 presents the numbers and distribution of actinomycetes as well as fungi, heterotrophic and autotrophic bacteria in different seasons. During the winter months (18-20°C), total bacterial counts were between 10^4 and 10^5 cfu/g of stone while actinomycetes ranged between 10^3 and 10^4 cfu/g; in summer, lower counts were recorded.

Table 3. Abundance of microorganisms on stones from a Tell Basta tomb during different seasons

Organism	Microbial count (cfu/g stone)			
	Winter		Summer	
	Sound	Decayed	Sound	Decayed
Bacteria	8100	30000	2100	7100
Actinomycete	90	2600	70	1700
Fungi	600	360	1500	580

The abundance of actinomycetes on decayed stone was always higher than on sound, similar to the findings for Portchester Castle. A total of 126 isolates have been characterised. The actinomycete isolates were assigned to 4 different taxonomic groups; 54% belonged to the *Streptomyces*-type group, 26% to the *Nocardia* group, 14% showed the characters of the *Micromonospora* group while the rest of the isolates analysed (6%) were assigned to the sporangiate-type group of actinomycetes. It was shown that most of the actinomycete isolates used a wide range of different carbon and nitrogen sources. About 53% of the strains studied were acid producers and 88% had the ability to produce extracellular pigments. Only 25% of the isolates studied showed tolerance to high salinity.

Thus filamentous bacteria appear to be very common on stone when moisture levels are very low, as for a hot English summer or the semi-arid climate of Egypt. The absence or lower numbers of filamentous bacteria on sound stone indicates that some degree of microbial succession occurs on stone. It is not clear whether actinomycetes on decayed stone are causative agents of decay or secondary colonisers after initial damage by other agencies. Mycelial forms are certainly more able to penetrate the substrate if it is friable and become established. However in short term laboratory weight loss studies, as described by Lewis et al. (1988), the isolated actinomycetes caused negligible weight loss. Under the SEM, however, there was clear evidence of pitting and erosion troughs around the margins of colonies attached to the stone disc surface but the effect was not enough to cause a measurable weight loss.

2.4 Biofilm

Examination of stone from monuments at Portchester, Khaniá and Petrás by electron microscopy showed that biofilms were always found on decayed stone. Biofilms on decayed stone at Portchester Castle were reported by Lewis (1987) and Tayler (1991) and they indicated a mixed microbial population embedded in a polymeric matrix. Fungi, algae, filamentous and unicellular bacteria were found in complex communities. Few organisms were visible on the surface and most colonisation was 3-4 mm into the stone. Hyphal strands of fungi extended into the stone via pores and these were in close association with unicellular bacteria, possibly as a consequence of a water coating on the strands. Further recent investigations of the biofilm on stone samples from the sheltered sites at Portchester showed that the biofilm is of a slime nature, probably due to the environmental conditions promoting the formation of a hydrophilic gel, which absorbs water from the air and releases it under conditions of low relative humidity (Warscheid et al., 1993). Where stone at Portchester was exposed to prevailing winds and marine spray, the biofilm was mostly heterogeneous and other particles of material were present in the biofilms which appeared contracted and flocculated. These particles have often been described by other workers as inorganic salts, carbonates of microbial origin or clay minerals (Thierry and Sand, 1995; Rodriguez-Navarro et al., 1997). These particles significantly increase the density and hydraulic properties of the EPS (Christensen and Characklis, 1990). Both stone monuments in Crete supported stone which had this type of biofilm and we believe that high salinity and other salts is reducing the viscosity of the EPS, similar to the flocculate mats observed by Gerdes et al. (1994).

The occurrence of biofilm on stone from Portchester and the monuments in Crete was investigated by measuring the total carbohydrates using the phenol/sulphuric acid colorimetric technique of Dubois et al. (1956), as modified by Lui et al. (1973) for lake sediments. This provides a rough and rapid estimation of carbohydrate derived from biofilm accumulated in stone specimens. Biofilms consist mostly of polysaccharides (Christensen and Characklis, 1990) and this method converts the polymers to free sugars which can be determined spectrophotometrically. Table 4 shows the results of surveys at 3 sites, with particular emphasis on Portchester Castle. This shows that the amounts varied from 2 to 113 μg/g stone with, in general, protected sites having higher levels than exposed sites.

Exposed substrates have been found to be subjected to constant material loss induced by rainwater (Biscontin et al., 1991) and deterioration due to frequent temperature and humidity variations (Halsey et al., 1998). It could also be speculated that sheltered sites supported higher numbers of bacteria as they might be richer in nutrients and thus more amenable for colonisation by biofilm microorganisms.

Table 4. Concentration of extractable carbohydrate in stones from UK and Cretan monuments

Monument	Site	Aspect	Total Carbobohydrate (μg/g stone)
Portchester Castle	IR	Exposed	19
	OT	Exposed	13
	WS	Protected	28
	GP	Protected	54
	BC	Protected	44
	S1	Exposed	75
	S2	Protected	113
	S3	Protected	40
	S5	Exposed	3
Khaniá walls	K3	Protected	32
Petrás Palace	P2	Exposed	2

2.5 Salt tolerance

The addition of NaCl to a basic heterotrophic medium has been used to assess the presence of halotolerant and moderate halophilic bacteria from wall paintings (Schostak and Krumbein, 1992). These types might also be expected to occur in the monuments in this study since they were found in coastal environments and the species present would be the ones best adapted to the environmental conditions. Investigations showed that a basic nutrient medium for heterotrophic bacteria (BR-11 mineral medium containing casein and glucose) supplemented with 0.09 M and 0.86 M NaCl gave the higher counts and these were selected for use in survey work. Further salt addition had an inhibitory effect on the recovery of bacteria.

Table 5 shows that the highest counts at Portchester and Khaniá were obtained using BR-11medium supplemented with 0.86 M NaCl. This was not the case for Petrás where a more variable picture emerged. In any event the results confirm that large numbers of bacteria exist on stone which can tolerate very high concentrations of a salt such as sodium chloride. We have also investigated the involvement of mixed populations of bacteria in wet/dry cycling of stone in a laboratory simulation in the presence of different salts. This will be published elsewhere but it confirms the field observations of extreme tolerance of bacteria to high salt levels.

Sodium chloride is thought to alter the microbial balance by inhibiting some species and enhancing others as well as influencing nutrient availability (Tresner and Hayes, 1970). Damage is enhanced in Petrás and Khaniá due to salt hydration and crystallisation. In Portchester, constant high external %RH and moderate temperatures did not lead to the macroscopic evidence of salt efflorescences on the surfaces of stones. The Mediterranean climate permits the existence of larger amount of surface salts which are washed away by the rain in the UK (Camuffo, 1995).

Table 5. Counts of bacteria obtained on BR-11 medium supplemented with different concentrations of sodium chloride

		Bacterial count (cfu/g stone)		
		BR-11 alone	+ 0.09 M NaCl	+ 0.86 M NaCl
Portchester	WS	70000	570000	470000
	S3	22000	10000	100000
Khaniá	K1	56000	18000	200000
	K2	2200	80000	220000
Petrás	P3	32000	96000	Nd
	P5	430000	48000	16000

In Crete, although the populations were not correlated with temperature or rainfall some relationship was found between location of the sampling and the bacterial counts. Since the weather conditions reach extremes, in terms of temperature and rainfall during the year, the biological factor predominated in the more sheltered, protected and less exposed areas. In Petrás, higher numbers of heterotrophic and acidophilic sulphur bacteria were also recovered from areas which faced south *i.e.* not directly exposed to the prevailing winds. In Khaniá, higher halotolerant and moderate halophilic bacterial populations were established on the sheltered substrates, which were not directly exposed to the seawater, wind and marine aerosols. This was also noted for the growth of plants.

3. CONCLUSIONS AND PERSPECTIVES

We have examined the bacterial populations that occur on stone monuments in temperate (UK), Mediterranean (Crete, Greece) and semi-arid (Egypt) regions in a series of studies over the last 8 years. In all cases, large populations of bacteria were observed by electron microscopy and these were often embedded in extensive sheets of protective biofilm. At Portchester Castle (UK) qualitative changes in bacterial population were observed and Gram-negative bacteria declined in numbers during the warmer summer months (16-20°C) when rainfall was at its lowest. Gram-positive types such as *Bacillus* became dominant. Throughout the study, many bacteria could not be sub-cultured and these were mostly associated with sound stone. Actinomycetes, filamentous bacteria belonging predominantly to the genus *Streptomyces*, were only found on decayed stone and could never be isolated from sound stone. Numbers of filamentous bacteria were highest during the winter and spring months and this corresponded to the period of greatest diversity. Actinomycetes caused little weight loss in short term studies but SEM studies revealed clear evidence of erosion troughs around colony margins on the stone surface.

In Crete, bacterial populations were investigated in building stone from the Minoan Palatial Building at Petrás and the fortified walls of Khaniá. Electron microscopy demonstrated the existence of extensive production of biofilms with bacteria embedded in them. The numbers and distribution of heterotrophic bacteria could not be related to seasonal changes in temperature and rainfall. In Khaniá, higher counts of halophilic and halotolerant heterotrophic bacteria were found in the sheltered areas which were not directly exposed to seawater and stone which showed evidence of other biological growths and salt efflorescences.

Water plays a deteriorative role in Portchester as the Castle constantly accepts rain throughout the year. Stone surface and structure are kept wet, making the penetration of the water through capillaries easier. Under Mediterranean climatic conditions, Petrás and Khaniá are left dry for long periods and water of a sudden rain cannot easily penetrate, as capillaries are not covered by water layers that would enhance circulation (Camuffo, 1995). Water causes mechanical alterations of stone (Nishioka and Harada, 1958) and takes part in all biological processes (Wiggins, 1990; Andersson et al., 1997). All three monuments are mechanically attacked by wind. Surface water is progressively displaced towards the interior of the stone while external temperature variations may increase the pressure of the entrapped air and the water vapour inside the cavities (Camuffo, 1995).

Under Mediterranean conditions at Khaniá and Petrás, the differences in the nature of bacterial populations appeared to be related to the location within the monument rather than to changes in the climate during the year. Under temperate climatic conditions at Portchester, actinomycetes were only found on decayed stone and could never be isolated from sound stone but were much more extensively distributed in stonework from Egypt. For all sites, stones of lower mechanical strength supported higher counts of heterotrophic bacteria. Biofilms were found extensively at all sites, from all locations, irrespective of climate.

Bacteria are thus widely distributed on stone experiencing seasonal changes in temperature and moisture and these are usually associated with the presence of biofilm, which provide a stable environment and may offer protection from osmotically unfavourable accumulations of salts. There is evidence to suggest that variations in the nature of bacterial populations may be dependent on season in temperate regions but more on location in semi-arid regions. Biofilm production and perhaps halotolerance provide means by which bacteria resist adverse changes in moisture levels on a temporal or spatial basis. Extreme changes in moisture may induce major shifts in the nature of bacterial populations so that filamentous, spore-forming types are selected which penetrate deeper into stone initiating more damage.

ACKNOWLEDGEMENTS

The authors are grateful for funding from the UK Building Research Establishment, Garston (ST), the Greek Public Benefit Foundation A. S. ONASSIS (SP), and the Egyptian Government (HA) for the different stages of the work. Dr M. Tsipopoulou (24th Archaeological Ephorate of Prehistoric and Classical Antiquities, Ag. Nikolaos, East Crete), Prof. Th. Markopoulos (Technical University of Khaniá, Crete) and Mr F. Powell (English Heritage) all facilitated sampling and provided information and scientific background on the study sites. FORTH-IMBB (Heraclion, Crete), and Prof. A. Economou generously provided hospitality and laboratory facilities for the part of the project in Crete. Professor Mahmoud Abd-El Razik (SCU) and Professor Mahmoud Omar (Zagazig University) helped to make work possible in Egypt. We thank Mr Derek Weights (Geology Department, University of Portsmouth, UK) for the XRD analysis of the building stones and Dr S. T. Moss (School of Biological Sciences, University of Portsmouth, UK) and Mr B. Loveridge (Geology Department, University of Portsmouth, UK) for their assistance with SEM.

REFERENCES

Andersson, M. A., Nikulin, M., Kõljalg, U., Andersson, M. C., Rainey, F., Reijula, K., Hintikka, E. -L. and Salkinoja-Salonen, M. 1997. Bacteria, moulds and toxins in water-damaged building materials. Appl. Environm. Microbiol. **63**: 387-393.

Appolonia, L., Vaudan, D. and De Leo, S. 1996. Projet des traveaux de conservation du théâtre romain d' Aoste: Recherche et résultats. *In* Proceedings of the 8th International Congress on Deterioration and Conservation of Stone vol. 2, J. Riederer (ed.), Berlin p 1097-1107.

Biscontin, G., Felix, C., Maravelaki, P. and Zendri, E.,1991. Characteristics, weathering forms and mechanisms of Istria stone in Venice. *In* Proceedings of the 2nd International Symposium on the Conservation of Monuments in the Mediterranean Basin D. Decrouez,

J. Chamay and F. Zezza (eds.), Ville de Genève, Musée d'Histoire Naturelle et Musée d'Art and d'Histoire p 141-153.

Bland W. and Rolls, D. 1998. Mechanical weathering processes. *In* Weathering. An Introduction to the Scientific Principles, Arnold, Great Britain p 85-114.

Camuffo, D. 1995. Physical weathering of stones. The Science of the Total Environment **167**: 1-14.

Christensen, B. E. and Characklis, W. G. 1990. Physical and chemical properties of biofilms *In* Biofilms W. G. Characklis and K. C. Marshall (eds.), Wiley-Interscience Publication p 93-130.

Decho, A. W. 1994. Molecular-scale events influencing the macroscale cohesiveness of exopolymers. *In* Biostabilisation of minerals W. E. Krumbein, D. M. Peterson and L. J. Stal (eds.), Springer-Verlag p 138-148.

Dubois, M., Gilles, K.A., Hamilton, J.K., Rebers, P.A. and Smith, F. 1956. Colorimetric method for determination of sugars and related substances. Anal. Chem. **28**: 350-356.

Eckhardt, F. E. W. 1988. Influence of culture media employed in studying microbial weathering of building stone and monuments by heterotrophic bacteria and fungi. *In* 6th International Congress on Deterioration and Conservation of Stone, N. Copernicus University, Torun, Institute of Conservation and Restoration of Cultural Property, Press Department p 71-81.

El-Nakeeb, M.A. and Lechevalier, H.A. 1962. Selective isolation of aerobic actinomycetes. J. Appl. Microbiol. **2**: 75-77.

Fookes, P. G., Gourley, C. S. and Ohikere, C. 1988. Rock weathering in engineering time. Quart. J. Eng. Geol. **21**: 33-57.

Friedmann, E.I. and Ocampa-Friedmann, R. 1984. Endolithic microorganisms in extreme dry environments: analysis of a lithobiontic habitat. *In* Current Perspectives in Microbial Ecology M.J. Klug and C.A. Reddy (eds.), Proceedings of the Third International Symposium on Microbial Ecology, Michigan State University p 177-185.

Gerdes, G., Dunajtschik-Piewak, K., Riege, H., Taher, A. G., Krumbein, W. E. and Reineck, H.-E. 1994. Structural diversity of biogenic carbonate particles in microbial mats. Sedimentology **41**: 1273-1294.

Guidobaldi, F. and Mecchi, A. M. 1985. Corrosion of marble by rain. The influence of surface roughness, rain intensity and additional washing. *In* Proceedings of the 5th International Congress on Deterioration and Conservation of Stone, vol. 1, Lausanne, Presses Polytechniques Romandes p 467-474.

Guillitte, O. 1995. Bioreceptivity: a new concept for building ecology studies. The Science of the Total Environment **167**: 215-220.

Halsey, D. P., Mitchell, D. J. and Dews, S. J. 1998. Influence of climatically induced cycles in physical weathering. Quart. J. Eng. Geol. **31**: 359-367.

Jager, G. and Bruins, E.H. 1974. Effect of repeated drying at different temperatures on soil organic matter decomposition and characteristics, and on the soil microflora. Soil Biol. Biochem. **7**: 153-159.

Jenneman, G. E., McInerney, M. J. and Knapp, R. M. 1985. Microbial penetration through nutrient-saturated Berea sandstone. Appl. Environm. Microbiol. **51**: 383-391.

Krumbein, W. E., Diakumaku, E., Gehrmann, C., Gorbushina, A. A., Grote, G., Heyn, C., Kuroczkin, J., Schostak, V., Sterflinger, K., Warscheid, T., Wolf, B., Wollenzien, U., Yun-Kyung, Y. and Petersen, K. 1996. Chemoorganotrophic microorganisms as agents of decay in the destruction of objects of art - a summary. *In* Proceedings of 8th International Congress on Deterioration and Conservation of Stone, vol. 2, J. Riederer (ed.), Berlin; p. 631-636.

Krumbein, W.E. and Gorbushina, G. 1995. Organic pollution and rock decay. *In* Biodeterioration of Constructional Materials, L.H.G. Morton (ed.) p. 277-284.

Kuster, E. and Williams, S.T. 1964. Selection of media for isolation of actinomycetes. Nature **202**: 928-929.

Lammel, G. and Metzig, G. 1997. Pollutant fluxes onto the facades of a historical monument. Atmospheric Environment **31**: 2249-2259.

Lappin-Scott, H. M. and Costerton, J. W. 1990. Starvation and penetration of bacteria in soils and rocks. Experientia **46**: 807-812.

Lewis, F.J. 1987. PhD thesis (CNAA) Investigations of bacteria on building stone and their role in stone decay. Portsmouth Polytechnic.

Lewis, F. J., May, E. and Bravery, A. F. 1988. Metabolic activities of bacteria isolated from building stone and their relationship to stone decay. *In* Biodeterioration 7 p. 107-112.

Lui, D., Wong, P.T.S. and Dutka, B.J. 1973. Determination of carbohydrate in lake sediment by a modified phenol-sulfuric acid method. Water Res. **7**: 741-746.

Mallory, L.M., Austin, B. and Colwell, R.R. 1977. Numerical taxonomy of bacteria isolated from the estuarine environment. Can. J. Microbiol. **23**: 733-750.

Nishioka, S. and Harada, T. 1958. Elongation of stones due to absorption of water. *In* Review of 12th Meeting of Japan Cement Association, Tokyo p. 66-67.

Palmer, R. J. Jr. and Hirsch, P. 1991. Photosynthesis-based microbial communities on two churches in northern Germany: weathering of granite and glazed brick. Geomicrobiol. J. **9**: 103-118.

Puhringer, J. 1996. Deterioration of materials by hydraulic pressure in salt/water systems - an outline model. *In* Proceedings of the 8th International Congress on Deterioration and Conservation of Stone, vol. 1, J. Riederer (ed.), Berlin p. 545-556.

Rodriquez-Navarro, C., Sebastian, E. and Rodriquez-Gallego, M. 1997. An urban model for dolomite precipitation: dolomite on weathered building stones. Sediment. Geol. **109**: 1-11.

Sand, W. 1997. Microbial mechanisms of deterioration of inorganic substrates a general mechanistic overview. Internat. Biodet. Biodeg. **40**: 183-190.

Schostak, V. and Krumbein, W. E. 1992. Occurrence of extremely halotolerant and moderate halophilic bacteria on salt impaired wallpaintings. *In* Proceedings of 7th International Congress on Deterioration and Conservation of Stone J. Delgado Rodriguez, F. Henriques and F. Telmo Jeremias (eds.), Lisbon p. 551-560.

Silva, B., Prieto, B., Rivas, T., Sanchez-Biezma, M. J., Paz, G. and Carbellal, R. 1997. Rapid biological colonisation of a granitic building by lichens. Internat. Biodet. **40** : 263-267.

Smith, D. I., Greenaway, M. A., Moses, C. and Spate, A. P. 1995. Limestone weathering in eastern Australia. Part 1: Erosion rates. Earth Surface Processes and Landforms **20**: 451-463.

Tayler, S. 1991. PhD thesis Composition and Activity of Bacterial Populations found on Decaying Stonework. University of Portsmouth, UK.

Tayler, S. and May, E. 1991. The seasonality of heterotrophic bacteria on sandstones from ancient monuments. Internat. Biodet. **28**: 49-64.

Tayler, S. and May, E. 1995. A comparison of methods for the measurement of microbial activity on stone. Studies in Conservation **40**: 163-170.

Thierry, D. and Sand, W. 1995. Microbially Induced Corrosion. *In* Corrosion Mechanisms in Theory and Practice, P. Marcus and J. Oudar (eds.), Series in Corrosion Technology, **8**: 457-498. Marcel Dekker Inc.

Tiano, P., Accolla, P. and Tomaselli., L. 1995. Phototrophic biodeteriogens on lithoid surfaces: an ecological study. Microb. Ecol. **29**: 299-309.

Tresner, H. D. and Hayes, J. A. 1970. Improved methodology for isolating soil microorganisms. Appl. Microbiol. **19**: 186-187.

Turk, N. and Dearman, W. R. 1986. Influence of water on engineering properties of weathered rocks. *In* Groundwater in Engineering Geology, Geological Society Engineering Geology Special Publication, no. 3, J. C. Cripps, F. G. Bell and M. G. Culshaw (eds.) p. 131-138.

Warke, P.A. and Smith, B.J. 1994. Inheritance effects on the efficacy of salt weathering mechanisms in thermally cycled granite blocks under laboratory and field conditions. *In* Granite Weathering and Conservation. E. Bell and T.P. Cooper (eds.), Proceedings of a conference at Trinity College, Dublin p. 19-27.

Warscheid, T. 1990. PhD thesis. Untersuchungen zur Biodeterioration von Sandsteinen unter besonderer Berücksichtigung der chemoheteroorganotrophen Bakterien. University of Oldenburg, Germany.

Warscheid, T., Becker, T. W., Braams, J., Gehrmann, C., Krumbein, W. E., Petersen, K. and Bruggerhoff, S. 1993. Studies of the temporal development of microbial infection of different types of sedimentary rocks and its effect on the alteration of the physico-chemical properties in building material. *In* Conservation of Stone and Other Materials, M. J. Thiel (ed.), Proceedings of the International RILEM/UNESCO Congress. vol. 1. Causes of Disorder and Diagnosis, Paris, France. UNESCO p. 303-310.

Webley, D.M., Henderson, M.E.K and Taylor, I.F. 1963. The microbiology of rocks and weathered stones. J. Soil Sci. **14**: 102-112.

Wiggins, P. M. 1990. Role of water in some biological processes. Microbiol. Rev. **54**: 432-449.

OCCURRENCE AND FLUCTUATION IN PHOTOSYNTHETIC BIOCOENOSES DWELLING ON STONE MONUMENTS

Luisa Tomaselli[1], Piero Tiano[2] and Gioia Lamenti [1]
[1]*Consiglio Nazionale della Ricerche - C. S. Microrganismi Autotrofi, P.le delle Cascine, 27, I-50144 Firenze, Italy;* [2]*Consiglio Nazionale della Ricerche - C. S. Cause di Deperimento e Conservazione delle Opere d'Arte, Via Alfani, 76, I-50121 Firenze, Italy.*

Key words: biodeterioration, stone monuments, phototrophic microorganisms.

Abstract: This research deals with the occurrence and fluctuation of phototrophic microbial communities dwelling on stone monuments. We examined many photosynthetic biocoenoses of several Italian monuments. The most widespread and recurring taxa occurring in the epilithic communities were among the cyanobacteria *Chroococcidiopsis*, *Pleurocapsa*, *Leptolyngbya* and *Plectonema*, and among the chlorophytes were *Coccomyxa*, *Apatococcus*, *Chlorococcum* and *Stichococcus*. The community composition was subjected to small seasonal changes affecting the amounts of the different components more than the qualitative biocoenosis structure. The number of phototrophic microorganisms increased slightly from late spring to autumn, in accordance with the improvement of climatic conditions. In winter and early spring, we found the lowest amount of phototrophic microorganisms. A detailed study, carried out on marble statues in the Boboli Garden (Florence, Italy), showed that the first coloniser was the green alga *Coccomyxa*. We observed that stone colonisation started about one year after restoration. The success of *Coccomyxa* in marble stone colonisation could be attributed to its specific growth strategy and features.

1. INTRODUCTION

In recent decades, the rates of stone decay processes have undergone dramatic acceleration owing to the increase of environmental pollution.

Monuments located outdoors are affected not only by physical and chemical weathering but also by the biological activities of stone-dwelling microorganisms, among which the phototrophs often prevail (Gómez-Alarcón et al., 1995; Urzì et al., 1994; Ortega-Calvo et al., 1993). The characterisation of these microorganisms and a clear understanding of their role in stone decay processes are important steps for planning the most suitable restoration interventions.

The aim of our research was to investigate the occurrence and fluctuation of phototrophic microbial communities dwelling on stone monuments in order to better understand their relationships with the lithotypes and the role of the specific factors that allow them to become established and flourish.

1.1 Investigated Monuments

One of the consequences of microbial development is the formation of unaesthetic pigmented biofilms, like those covering the statue shown in Figure 1.

We examined many photosynthetic biocoenoses occurring on several Italian monuments. Table 1 reports the investigated monuments.

Table 1. Investigated monuments

Monument	Environment	Lithotype
1 Roman Statue (Volterra, SI)	Urban area, low humidity, high light intensity	Carrara marble
2 Cathedral (SI)	Urban area; low humidity, high light intensity	Montagnola marble
3 Crypt, Church M.Favana (LE)	Hypogean environment, high humidity	Bioclastic limestone
4 Leaning Tower (PI)	Urban area (grassy)	St. Giuliano marble
5 Medici Fortress (LI)	Port area, marine aerosol, high light intensity	Brick
6 The Pyramid (FI)	Urban park, high humidity, low light intensity	Sandstone
7 Tacca's Fountains (FI)	Urban area, running water, high light intensity	Carrara marble

A detailed study was carried out on some marble statues in the Boboli Garden (Florence, Italy) (Tab. 2), which we used as an investigative model for stone monument biodeterioration.

We monitored the fluctuation of the colonising process over a 3 year period. We studied statues that had not been restored recently and statues that had been restored at different times, from 2 months to 6 years prior to the study (Figs 2, 3). Examination of the biocoenoses, occurring on some recently-restored statues, allowed us to assess also the dynamics of the colonising process.

Figure 1. Musicante, statue not recently restored.

Table 2. The Boboli Garden investigated marble statues

Statue	Characteristics
1. Calliope	restored 2 months before sampling
2. Ganimede	restored 12 months before sampling
3. Giove Olimpico	restored 24 months before sampling
4. Abbondanza	restored 6 years before sampling
5. Flora	Not restored
6. Musicante	Not restored
7. Contadino	Not restored
8. Zappatore	Not restored

Analysis of the occurrence and distribution of the phototrophic microorganisms was performed using light and epifluorescence microscopy and specific cultural media. The Most Probable Number (MPN) method was used for evaluating phototrophic microbe counts (Lamenti et al., 1998). We isolated the most representative photosynthetic microorganisms to determine the nature and composition of the microbial communities; we then identified them at the generic level and characterised the strains obtained in axenic cultures.

Figure 2. Giove Olimpico: marble bust restored 24 months before the first sampling.
Figure 3. Ganimede: marble statue restored 12 months before the first sampling.

2. STRUCTURAL FEATURES OF THE BIOCOENOSES

Light and epifluorescence microscopy observations showed an abundance of phototrophic microorganisms in the epilithic microbial communities (Fig 4).

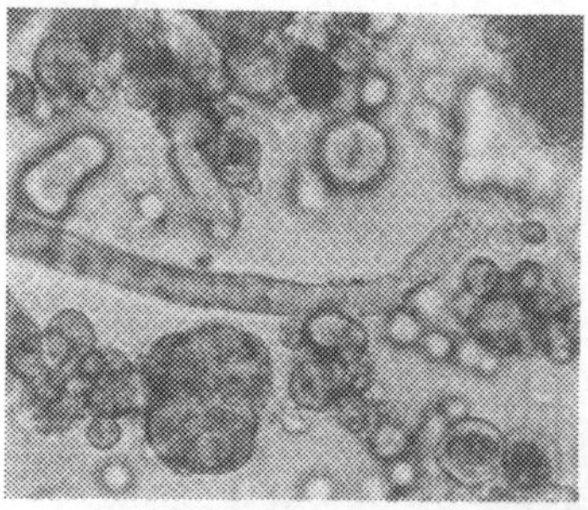

Figure 4. Micrograph of photosynthetic biocoenosis occurring on the Musicante statue (x 400).

These microbial communities form thick biofilms, with intense pigmentation varying from dark-green to dark red that considerably alters the aesthetic appearance of the monuments. On recently-restored statues, we found the development of these biofilms to be less extensive and also of lesser thickness. Enumeration of total epilithic phototrophs showed values ranging from 10^3 to 10^5 cells per cm^2 of surface.

Tables 3 and 4 report the genera most commonly occurring in the samples we investigated.

Table 3. Occurrence of phototrophic microorganisms in the samples taken from the stone monuments reported in Table1.

	Monument						
	1	2	3	4	5	6	7
CYANOBACTERIA							
Synechococcus				+			
Gloeocapsa							
Chroococcidiopsis				+	+	+	+
Myxosarcina		+			+		
Pleurocapsa		+		+		+	
Geitlerinema	+	+					+
Leptolyngbya				+	+		
Plectonema			+			+	
Nostoc		+					
Scytonema			+				
CHLOROPHYTA							
Apatococcus		+					+
Chlorella		+					+
Chlorococcum						+	+
Coccomyxa							
Rhizothallus							
Stichococcus	+	+		+			
Trentepohlia	+	+					
Ulothrix					+	+	
BACILLARIOPHYTA							+

The photosynthetic communities were constituted prevalently by cyanobacteria with a few green microalgae. Unicellular cyanobacteria showed various morphotypes, often occurring in cell aggregates enclosed by thick, sometimes coloured, sheaths; this feature was observed quite commonly in samples taken during the summer. Filamentous forms formed a dense network with entangled cells of the coccoid forms.

In some cases, such as the crypt of the church M. Favana, near Lecce, we found only cyanobacteria, whereas on stones wetted by running water, as in the Fountains of Tacca, we found a rich epilithic community including many diatoms, in addition to chlorophytes and cyanobacteria. The most widespread and recurring taxa among the cyanobacteria were *Chroococcidiopsis*, *Pleurocapsa*, *Leptolyngbya* and *Plectonema*, and, among the chlorophytes, *Coccomyxa*, *Apatococcus*, *Chlorococcum* and *Stichococcus*.

Table 4. Occurrence of phototrophic microorganisms in the samples taken from the Boboli Garden statues reported in Table 2.

	Boboli Garden Statue							
	1	2	3	4	5	6	7	8
CYANOBACTERIA								
Synechococcus	+							
Gloeocapsa					+	+		
Chroococcidiopsis						+	+	
Myxosarcina				+				
Pleurocapsa	+		+			+		+
Geitlerinema							+	
Leptolyngbya	+		+		+			+
Plectonema				+		+		
Nostoc		+		+				
Scytonema						+		
CHLOROPHYTA								
Apatococcus				+	+			
Chlorella				+				
Chlorococcum				+	+			
Coccomyxa	+	+	+	+	+	+		+
Rhizothallus			+					
Stichococcus				+				

On the unrestored statues in the Boboli Garden, we found both filamentous and unicellular cyanobacteria and some genera of green microalgae. On the very recently-restored statues (from 2 months to about 10 months prior to sampling, e.g. Calliope), we did not find any phototrophic microorganisms. On statues restored less recently (i.e., more than 12 months prior to sampling), we detected a strong prevalence of green microalgae that formed a very thin film, constituted above all by *Coccomyxa* cells. On these statues, cyanobacterial species occurred in very small amounts and included only a few pleurocapsalean morphotypes. The most probable source of infection of marble statues by *Coccomyxa* are trees, on which this microalga is commonly found.

Samples taken from more recently-restored statues indicate that the establishment of colonies of green microalgae requires several months: in fact, they appeared about one year after the cleaning treatment. In any case, the first phototrophic coloniser was the green microalga *Coccomyxa,* which showed a marked increase during autumn (Fig. 5).

The composition of the communities on unrestored statues was subjected to small seasonal changes that affected the amounts of the different components rather than changing the qualitative structure of the biocoenosis. The number of phototrophic microorganisms increased slightly from late spring to autumn, in accordance with improved climatic conditions. We found the lowest amount of phototrophic microorganisms in winter and early spring. The relatively low seasonal fluctuations of the total number of phototrophs within the biocoenosis on unrestored statues has led us to consider these communities as stable and mature.

By pooling the data obtained from examination of recently-restored statues, we were able to reconstruct the dynamics of the colonising process (Fig. 6).

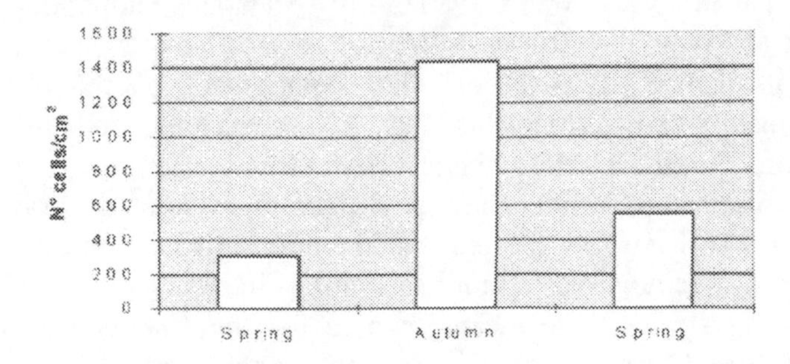

Figure 5. Seasonal fluctuation of total epilithic phototrophs occurring in the biocoenosis of recently-restored statues in the Boboli Garden (Florence, Italy).

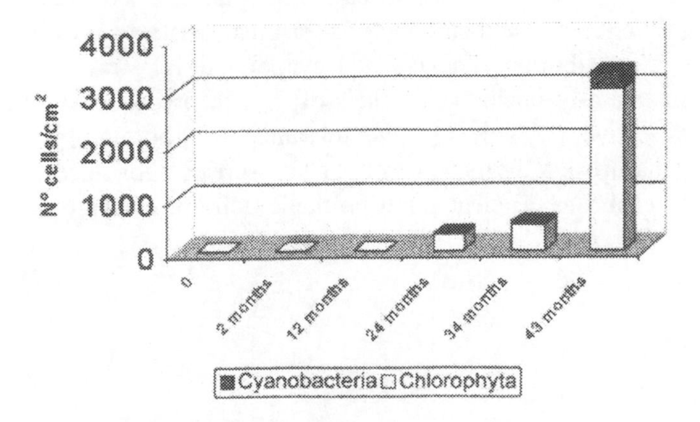

Figure 6. Development of epilithic phototrophic communities on recently-restored statues in the Boboli Garden (Florence, Italy).

3. SUBSTRATE FEATURES AND ENVIRONMENTAL CONDITIONS

Previous data showed that the majority of cyanobacteria preferred calcareous substrata like the Carrara marble statues that we have examined in the Boboli Garden. High stone porosity and rough surface played a greater role than did mineral composition in promoting microbial establishment. Microbial colonisation of calcareous stone surfaces was found to depend primarily on specific physical stone parameters and, of course, on environmental factors (Tiano et al., 1995). The most important of the environmental factors is water availability. Adequate temperature and solar irradiance, and type of atmospheric deposition are also relevant factors.

Solar irradiance affected both the development of the photosynthetic microorganisms constituting this type of ecosystem, and its qualitative composition. Plotting photosynthetic activity rates (P), expressed as rate of oxygen evolution per unit chlorophyll for the isolated strains, against irradiance (I) (P/I curves) showed that the chlorophycean strains had higher photosynthetic activity rates and saturation intensities (I_k) than did the cyanobacterial strains. The mean I_k value for the *Coccomyxa* strains was above 100 µmol/photons m^{-2} s^{-1}, and only about half this value for the cyanobacterial strains. The relative P values showed a similar trend (Fig 7). This behaviour explains the predominance of the green microalgae over the cyanobacteria on stone surfaces exposed to high photon flux densities (PFDs). The lower tolerance of the cyanobacterial strains justifies both their epilithic and endolithic distribution. The isolated cyanobacterial strains were able to maintain substantially unchanged values of P_{max} in presence of PFDs higher than the respective I_k. This relative tolerance to photoinhibition may be enhanced in nature by the production of sun-screen substances, which can contribute to ensuring efficient photosynthetic activity even under high PFD conditions.

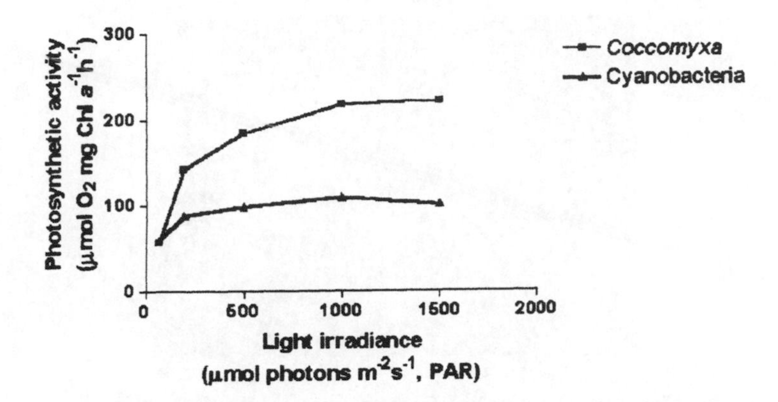

Figure 7. Mean photosynthetic activity values of four *Coccomyxa* and six cyanobacteria strains.

The higher photosynthetic activity of the isolated *Coccomyxa* strains was confirmed by their higher and faster growth rates in comparison to the cyanobacterial strains (Fig. 8).

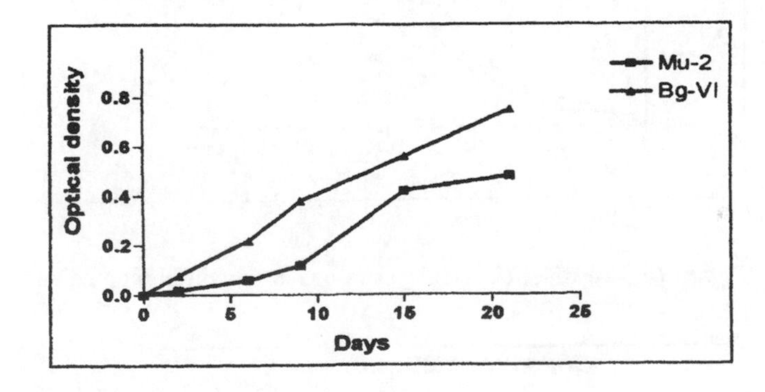

Figure 8. Growth of *Chroococcidiopsis* strain Mu-2 and *Coccomyxa* strain Bg-VI in liquid cultures.

4. INTRINSIC PROPERTIES

The main intrinsic properties responsible for the permanent establishment of microorganism colonies on stone are the capacity of adhesion, oligotrophy, metabolic flexibility and tolerance to adverse conditions.

All the isolated *Coccomyxa* strains showed to be facultative oligotrophs (Fig. 9). They were able to grow in very poor media (Fig. 10), unlike the cyanobacteria we tested (Fig. 11).

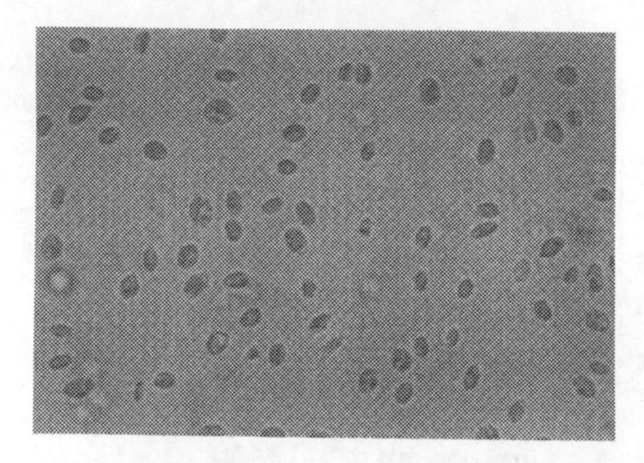

Figure 9. Micrograph of *Coccomyxa* strain Bg-VI (x 400).

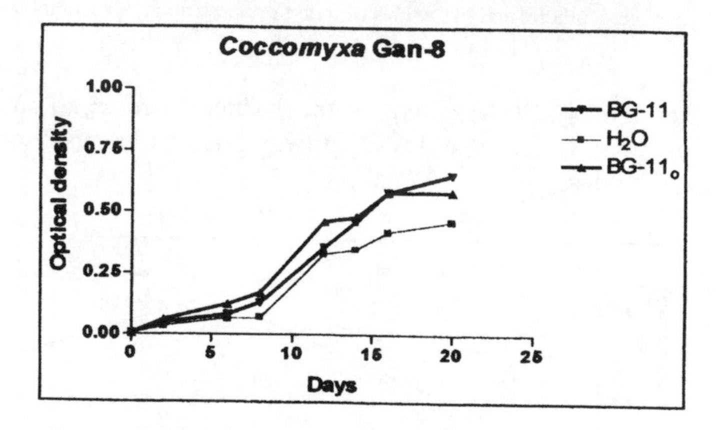

Figure 10. Growth of *Coccomyxa* strain Gan-8 in different media.

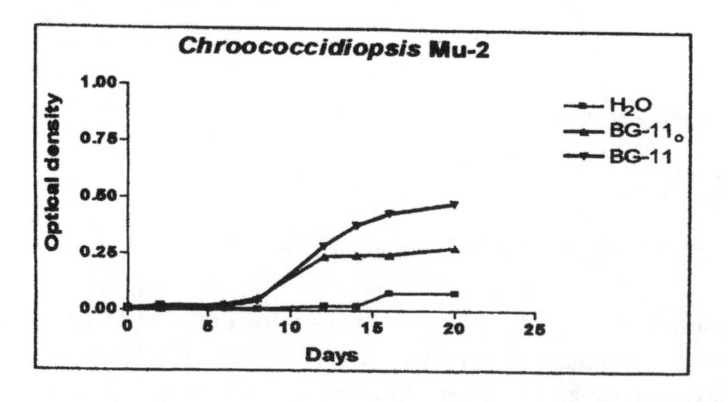

Figure 11. Growth of *Chroococcidiopsis* strain Mu-2 in different media.

Oligotrophy allows for initial substrate colonisation without immediate utilisation of the substrate, although this may follow once the organism has become established.

An important factor in the adhesion and proliferation of microorganisms on stone surfaces is the hydrophobicity of their cell surface. By using the biphasic water-hydrocarbon test system (Fattom and Shilo, 1984), we observed that cells of the *Coccomyxa* strains are completely concentrated in the hydrocarbon phase, and show a very high hydrophobicity (Fig. 12).

Microscopic observation showed that the hydrocarbon microdroplets were closely attached to the *Coccomyxa* cells; this indicates that the hydrophobic sites were distributed over the entire cell surface.

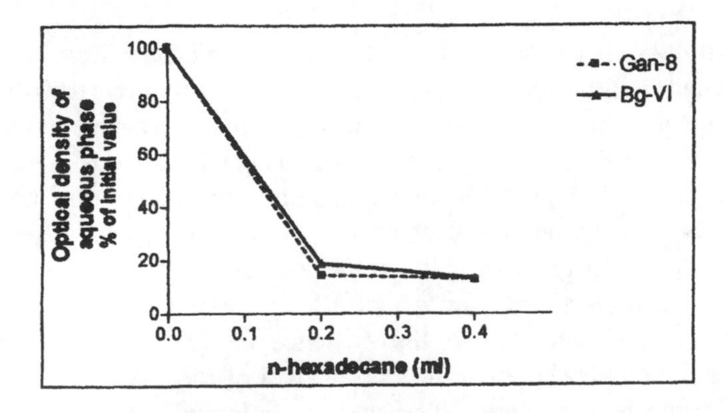

Figure 12. Partitioning of *Coccomyxa* strains Bg-VI and Gan-8 in the biphasic test system obtained by adding different amounts of *n*-hexadecane to 5 ml of cell suspension.

The isolated cyanobacterial strains did not show such a property and remained in the aqueous phase. The addition of cations to neutralise negative charges, commonly present on the cell envelope, and to allow the expression of hydrophobicity, did not cause any change of cell surface properties of the cyanobacterial strains tested. Only after removal of the external cell envelope by washing and addition of divalent cations did about 30% of cyanobacterial strains, of the genera *Chroococcidiopsis*, *Pleurocapsa*, *Nostoc* and *Plectonema*, change their behaviour and became biphasic. These strains showed a reduced and variable partitioning of the cells in the hydrocarbon phases, indicating low degrees of cell surface hydrophobicity.

About 70% of the *Chroococcidiopsis* cells remained in the aqueous phase (Fig. 13), while in *Pleurocapsa* this amount was reduced to 40%. Filamentous isolates of *Nostoc* and *Plectonema* showed a behaviour similar to that of *Pleurocapsa*.

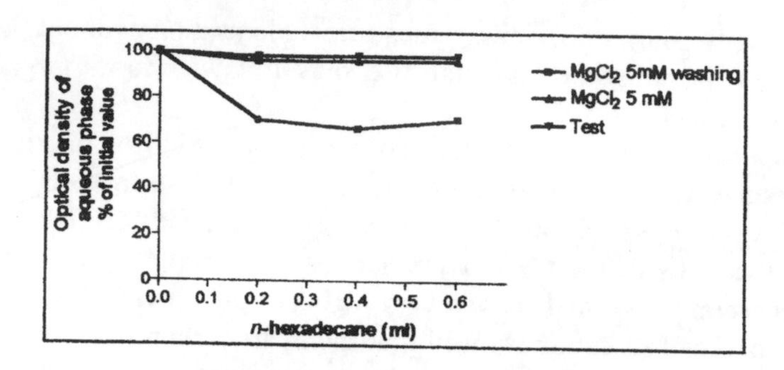

Figure 13. Partitioning of *Chroococcidiopsis* strain Mu-c in the biphasic test system with different amount of *n*-hexadecane. Effects of culture washing and cation addition.

Cell hydrophobicity was not the sole mechanism determining the adhesion of the isolated epilithic microorganisms. Excreted polymeric substances like sheaths, capsules and slimes, even if not directly observable under a light microscope, were detected in several cyanobacterial strains after negative staining (Fig. 14). About 90% of the isolated cyanobacterial strains were found to secrete extracellular substances. These substances were primarily polysaccharides with reactive acid groups on the carbohydrate residues, as revealed by specific staining with Alcian Blue (Crayton, 1982). The polymers become highly hydrated due to the presence of acid residues. The water retention capacity, besides ensuring the survival of the sheathed organisms, is involved in mechanical stone disgregation following alternate periods of hydration and dehydration. The extracellular substances, in addition to joining organism and substrate, participate in the formation of microbial aggregates, which contribute to the development of firm biofilms.

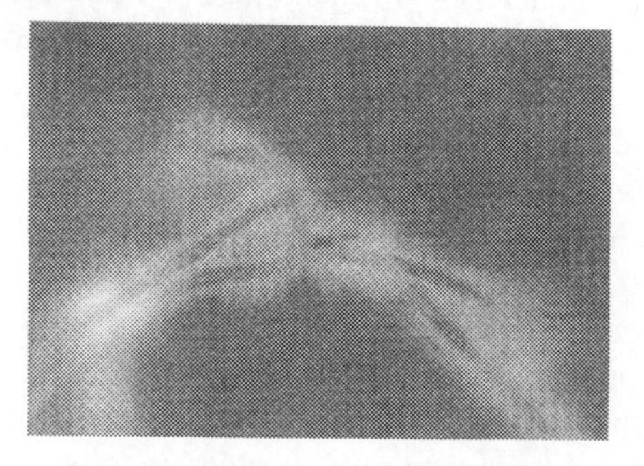

Figure 14. Micrograph of *Plectonema* Mu-pl strain showing the secreted exopolysaccharidic sheath (negative staining) (x 400).

5. CONCLUSIONS

The specific growth strategy of *Coccomyxa*, capable of facultative oligotrophy by scavenging nutrients from the environment, the high hydrophobicity of its cells and its tolerance to elevated PFDs, explain the success of this organism as first coloniser of the Boboli Garden marble statues. The observed abundant secretion of large and firm polysaccharidic substances by the isolated cyanobacterial strains justifies the establishment, proliferation and survival of these microorganisms in adverse conditions.

Taxonomic diversity among the photosynthetic microorganisms of this particular environment appears not to be very wide. This can be related to abiotic stresses. We found 10 taxa of cyanobacteria and 7 of chlorophyta. However, we are well aware that we studied only a part of the photosynthetic population, since many of the organisms present are unculturable. Our results are nevertheless in agreement with the data of other authors (Gómez-Alarcón et al., 1995; Ortega-Calvo et al., 1993).

6. PERSPECTIVES

Phenotypic and genotypic characterisation of the isolated representative strains constitute a starting point for the development of methods that allow the recognition of these organisms in natural habitats and the study of microbial diversity.

The difficulties related to the taxonomy of these phototrophs lead us to apply molecular techniques to investigate genetic biodiversity within homogenuous groups of cyanobacteria, like the *Pleurocapsa* group (Lamenti et al., 1999).

ACKNOWLEDGMENTS

Research supported by National Research Council of Italy, Special Project Cultural Heritage. Authors are grateful to Dr Litta Medri, Superintendent of the Sculptural Fittings of the Boboli Garden, Florence (Italy) and to Mr Domenico Mannelli of the DiSTAM of the University of Florence for his help in producing the graphic images.

REFERENCES

Crayton, M.A. 1982. A comparative study of volvocacean matrix polysaccharides. J. Phycol. **18**: 336-344.

Fattom, A. and M. Shilo. 1984. Hydrophobicity as an adhesion mechanism of benthic cyanobacteria. Appl. Environ. Microbiol. **47**: 135-143.

Gómez-Alarcón, G., M. Muñoz, X. Ariño and J.J. Ortega-Calvo. 1995. Microbial communities in weathered sandstones: the case of Carrascosa del Campo church, Spain. Sci. Total Environ. **167**: 249-254.

Lamenti, G., L. Tomaselli and P. Tiano. 1998. Successione microbica su statue di marmo. *Atti Convegno Congiunto ABCD, AGI, SIBBM, SIMGBM* 1-4 Ottobre 1998, Montesilvano Lido (PE), Italy.

Lamenti, G., M. Bosco , P. Tiano and L. Tomaselli. 1999. Biodiversity and molecular analysis of cyanobacteria from biofilms degrading works of art in Tuscany. Of Microbes and Art. The Role of Microbial Communities in the Degradation and Protection of Cultural Heritage. International Conference on Microbiology and Conservation, Florence, June 1999 p. 35-40.

Ortega-Calvo, J.J., P.M. Sanchez-Castillo, M. Hernandez-Marine and C. Saiz-Jimenez. 1993. Isolation and characterization of epilithic chlorophytes and cyanobacteria from two Spanish cathedrals (Salamanca and Toledo). Nova Hedwigia **57**: 239-253.

Tiano, P., P. Accolla and L. Tomaselli. 1995. Phototrophic biodeteriogens on lithoid surface: An ecological study. Microb. Ecol. **29**: 299-309.

Urzì, C. and W.E. Krumbein. 1994. Microbiological Impacts on the Cultural Heritage. *In* Durability and Change: The Science, Responsibility, and Cost of Sustaining Cultural Heritage, W.E. Krumbein, P. Brimblecombe, D.E. Cosgrove and S. Staniforth (eds.), Wiley and Sons Ltd, Chichester p. 107-135.

MICROBIAL COMMUNITIES IN SALT EFFLORESCENCES

Leonilla Laiz, Delfina Recio, Bernardo Hermosin and Cesareo Saiz-Jimenez
Instituto de Recursos Naturales y Agrobiologia, C.S.I.C., Apartado 1052, E-41080 Sevilla, Spain

Key words: efflorescences, halotolerant bacteria, *Bacillus*

Abstract: The weathering of stone and masonry results, among other phenomena, in the formation of efflorescences, a deposit of salts not originally evident in the sound material, and composed of a variety of different hygroscopic salts, including carbonates, chlorides, nitrates, sulphates, etc. Thus the deteriorated areas of some monuments can be considered as extremely saline environments, at least in the efflorescences zones. These zones seemed to be a particular niche for investigating halotolerant and/or halophilic bacteria, but only recently a few reports appeared stressing the importance of these types of bacteria. In this paper, bacteria were isolated from samples taken from the efflorescences originated in the Chapel of All Souls, cathedral of Jerez, Spain, and cultured in media with increasing concentrations of halite or epsomite (up to 25 %). The most frequently isolated genus was *Bacillus*, followed by *Staphylococcus*, *Kocuria*, *Micrococcus*, *Paenibacillus* and *Arthrobacter*. The origin, occurrence and ecology of halotolerant and halophilic bacteria in monuments are discussed.

1. INTRODUCTION

Weathering of exposed stone and masonry results from physical, chemical and biological processes, which includes dissolution of carbonates and sulphates, solubilization by leaching of elements from silicates, weathering due to crystallization and hydration pressures, microbial attack by inorganic and organic acids, etc.

Practically all building walls contain soluble salts, dispersed within the porous materials or locally concentrated. These salts are solubilized and migrate with the water in and out of the stone. The drying out of the solution at the exposed surface results in the formation of efflorescences. The term efflorescence is employed to denote the presence on stone or masonry surfaces of a deposit of inorganic substances not originally evident in the sound material. A description of the different types of efflorescences and the mechanisms of formation on building stones was reported by Charola and Lewin (1979).

Thus, a variety of different hygroscopic salts, including carbonates, chlorides, nitrates, sulphates, etc. can be found on the surfaces of decayed monuments. This means that some particular monument areas can be considered as extremely saline environments, at least in the efflorescences zones, but the salinity is not necessarily based on halite, as sometimes epsomite or gypsum are the main components.

Bacteria have adapted to growth in many different niches. One of them, the hypersaline environment is characterized by high concentrations of salts. Usually this refers to waters and the term hypersaline waters defines those which have concentrations of salts higher than seawater. Considerable research has been carried out on halophilic bacteria in relation to these environments (Rodriguez-Valera, 1988a; Javor, 1989; Vreeland and Hochstein, 1992). Comparatively, very little information exists for terrestrial environments.

In high saline niches, halotolerant bacteria and *Archaea* can thrive. Until recently *Archaea* were known only as inhabiting hostile environments (hot springs, solar ponds and saturated brines) where there is little microbial competition (Vreeland and Hochstein, 1992). To this fact also contributed that laboratory cultivation rarely included media with high salt concentrations and as a consequence this group tended to be overlooked.

Halotolerant bacteria, those which grow better without significant amounts of halite in their media, but can also grow at concentrations higher than those of seawater, can therefore be present in active form in saline environments and efflorescences (Incerti et al., 1997). Although efflorescences have a chemical origin, due to migration and crystallization of soluble salts (Mora et al., 1977), a biological origin, due to bacterial growth, was suggested by Lazar (1971) and Bassi and Giacobini (1973).

Lazar (1971) reported that large zones of the nave and altar paintings from the Cozia monastery (Romania) were covered by a whitish fairly adherent powder. The painting zones affected by the whitening phenomenon, which usually completely masked the painting was populated by numerous bacteria, as demonstrated both microscopic preparations of the powder scrapped off from the surface and inoculations on culture media. Of 94 isolates, the most abundant genera were *Bacillus*, *Arthrobacter*, and *Micrococcus*. Sporogenous forms belonging to various species of the genus *Bacillus* predominated (32 % of isolates). Almost half of the tested isolates reproduced the phenomenon when inoculated as culture suspensions on undamaged zones of the paintings, the most active being *Bacillus pumilus*, *Micrococcus roseus* and *Arthrobacter* sp. The powdery whitening was a consequence of the introduction of central heating into the monastery and the microclimate modification.

Further, Lazar and Dumitru (1973) reported that the frescoes affected by the efflorescence phenomenon from 10 monasteries of northern Moldavia were populated by a rich bacterial flora made up of representatives of the genera *Bacillus*, *Pseudomonas* and *Arthrobacter*. Of 117 isolates, 60 were *Bacillus* species (51%), 17 *Arthrobacter*, (15%) and 16 *Pseudomonas* (14%). As the authors previously demonstrated the capacity of bacteria to produce efflorescences, they considered that these genera should not be overlooked in the explanation of the causes of the blanching phenomenon of the frescoes.

Saiz-Jimenez and Samson (1981) studied the efflorescences of the frescoes of the monastery of Santa Maria de la Rabida (Spain), and identified gypsum as the most abundant salt, with minor amounts of calcium chloride, sodium chloride, potassium chloride, quartz and silicates. The most abundant bacterium was *Micrococcus luteus*, but species of the genera *Bacillus*, *Pseudomonas*, and *Streptomyces* were also identified. Sulphur oxiding and reducing bacteria in amounts exceeding those indicating biological attack to monuments (Tiano et al., 1975) were recorded.

In a recent study, Incerti et al. (1997) investigated the bacteria present in the efflorescences from a calcarenite monument. More than 40 % of the isolates from cultures with high salt concentrations were species of the genus *Bacillus*.

Although the presence of bacteria in mural paintings efflorescences was reported more than 25 years ago, it was only recently that efflorescences (Incerti et al., 1997) and deteriorated mural paintings (Rölleke et al., 1996) were considered as a potential interesting habitat for *Archaea*. Until now no cultivated halophiles were isolated from monuments, although, recently, Rölleke et al. (1998) detected and identified an *Archaea*l community within the deteriorated 13th century mural paintings in the chapel of the Herberstein Castle (Austria). The authors used denaturing gradient gel electrophoresis (DGGE) of PCR-amplified DNA encoding 16S rRNA. Total DNA was extracted from the wall painting material and the DNA fragments amplified with *Archaea*l specific primers. Six out of ten analysed samples showed the presence of *Archaea*l 16S rDNA fragments, identified as members or close relatives of the genus *Halobacterium*. This has important implications in the general ecological significance of *Archaea* on monuments with an considerable production of efflorescences.

The present study investigates the ecology of halotolerant and/or halophilic bacteria present in the deteriorated Chapel of All Souls in the Jerez Cathedral.

2. THE CATHEDRAL OF JEREZ

The cathedral of Jerez was constructed, with calcarenite from the quarries of Puerto de Santa Maria, between 1695 and 1778, when it was inaugurated. However, other works were carried out from 1780 onward, such as the construction of the Sacristy in 1783, and between 1788 and 1808 the construction of the Ciborium Chapel, anteroom to the Sacristy, Chapter Hall and Library. The construction progressed slowly until 1849 when work was definitively discontinued. In the present century restoration works were carried out in 1964, in the 70's, and between 1983-1986.

As a consequence of the studies carried out during the last restoration period, Rodriguez-Gordillo et al. (1988) reported that the calcarenite showed alteration phenomena in the west facing parts, with deep cavities in the ashlars. Other, even more obvious alteration phenomena were the mechanical fractures in the stone of the dome due to the pressure of the iron anchors of the railings, and the serious powdering of the calcarenite.

In addition, along the time, the east façade was markedly affected by dampness and the stone reredos of the Chapel of All Souls (now Chapel of the Resurrection) had a snowy appearance due to white efflorescences. Although in present times these are periodically cleaned off, it reappears.

Recent analysis of the efflorescences by X-ray diffraction revealed that the main component is hexahydrite ($MgSO_4.6H_2O$), accompanied by calcite, gypsum and quartz (Incerti et al., 1997). There are several hydrated phases of magnesium sulphate. The first one, precipitating from a saturated solution and at temperatures below 60°C, is epsomite ($MgSO_4.7H_2O$). Dehydration of epsomite originates hexahydrite. The predominance of one over other depends on the relative humidity. The hexahydrite identified in the studied samples arise from epsomite dehydration, which could be produced in the cathedral walls or after sampling and transport to the laboratory. However, epsomite was the main component of the efflorescences in the sampling carried out in 1983 (Rodriguez-Gordillo et al., 1988).

About 75 % of the efflorescence powder is soluble in water, due mainly to hydrated magnesium sulphate. Ion chromatography revealed that, in addition, nitrates amounting to 0.5% and chlorides to 0.3% are present. The X-ray analysis of the insoluble residue indicated the presence of calcite, quartz and poorly crystallized hydromagnesite [$Mg_5(CO_3)_4(OH)_2.4H_2O$]. Infrared spectrum of the residue matched those of artificial hydromagnesite (Van der Marel and Beutelspacher, 1976).

Hydromagnesite is recognized as the final-phase mineral in a Mg-rich mineral sequence in many caves and its formation related to Mg-enriched solutions and low P_{CO2} values. These conditions are consistent with highly alkaline waters and high evaporation conditions (Cañaveras et al., 1999). Hydromagnesite crystals were found in Altamira Cave partly embedded in biofilms, and microbial activity was suggested to be involved in the precipitation of hydromagnesite deposits.

In a previous study, several samples of mortars were analysed and yielded, as main components, gypsum, calcite, quartz and lower amounts of epsomite. Concerning the possible origins of the epsomite, the quarry stone had values of 0.02-0.04% of sulphates (MgO was 0.5%), and the Pliocene bedrock supporting the cathedral foundation contained 0.02% sulphates (Rodriguez-Gordillo et al., 1988).

Both the narrow street, the east-facing orientation of the external Chapel walls and the fact that the cathedral is built on the bank of a stream contribute to the problem of dampness, although the origin of epsomite should probably be found in the composition of the mortar, with a high amount of magnesium salts. The joints of the ashlars, made with mortar, are completely impregnated with dampness due to the hygroscopicity of the soluble salts, evidencing a clear liquid film. The high solubility of magnesium sulphate means that it is readily dissolved in the processes of water recirculation inside the stone and mortar, through the walls of the building.

3. ISOLATION TECHNIQUES AND CULTURE CONDITIONS

Efflorescences from the right side of the Chapel of All Souls were collected in sterile tubes. One g of each sample was added to distilled water and cultured in media with increasing concentrations of halite or epsomite (up to 25%). Culture media used in this study were previously described (Rodriguez-Varela et al., 1981).

4. BACTERIAL IDENTIFICATION

Total cellular fatty acid methyl esters (FAME) were analyzed using the MIDI system in accordance with the protocols for cultures grown on solid medium and instrument specifications recommended by Microbial

Identification System, Neward, USA. By using this commerciallly available gas chromatograph-software system, we obtained FAME profiles which were identified automatically by comparison with the Sherlock Standard Aerobe database. The dendrogram programme was used to determine FAME similarity among groups of bacteria (Gonzalez et al., 1999).

5. RESULTS AND DISCUSSION

The Chapel of All Souls has a long history of continuous production of epsomite efflorescences, the higher amounts being observed in the year 1983. Subsequently, the efflorescences were removed periodically in a routine cleaning process. The existence for long time of magnesium salts and their hygroscopicity undoubtedly could have selected a specific community adapted to growing in this high saline environment and it was expected that the bacteria present in efflorescences should mainly be halotolerant bacteria, with a lower, if any, representation of halophilic bacteria.

The existence of a considerable number of bacteria in efflorescences of monuments was already reported by Saiz-Jimenez (1982), the colony forming units (cfu) amounting up to 2×10^6 cfu/g of sample in the frescoes of the monastery of Santa Maria de la Rabida (Spain).

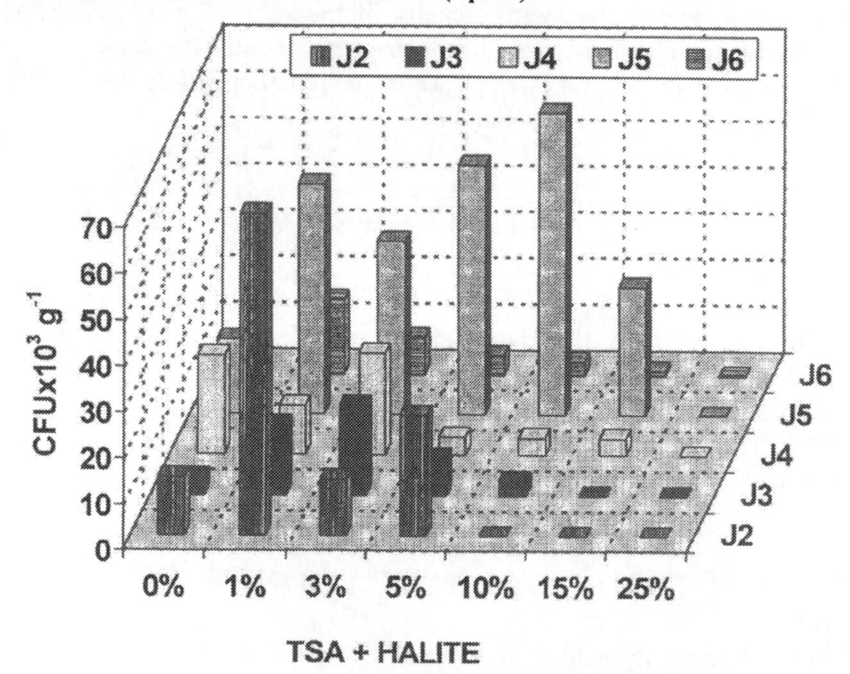

Figure 1. Colony forming units in samples inoculated in TSA+ halite

In the chapel of the cathedral the bacterial population ranged from 3.8 to 21.6 x 10^3 cfu/g in culture media without halite, measured after 72 h of incubation. Salinity concentrations from 0.9 to 15% of halite generally increased the number of cfu with respect to medium without halite (54.3 to 70.7 x 10^3 cfu/g). However, the highest salinity concentrations reduced the population, except for sample J-5, in which the population obtained with 10 % halite was higher (65.6 x 10^3 cfu/g) than for other ranges (Figure 1). Interestingly, in this sample, the lowest amount was obtained in the medium without halite. Estimation of the population present in the media with increasing concentrations of epsomite showed similar trends. In this case, the highest numbers were obtained for sample J-3 at 10% (33.6 x 10^3 cfu/g) and J-5 at 5% epsomite (48.3 x 10^3 cfu/g) as shown in Figure 2. The estimation of the population growing with epsomite was considered necessary and complementary to the estimation with halite because of the specific composition of the efflorescences.

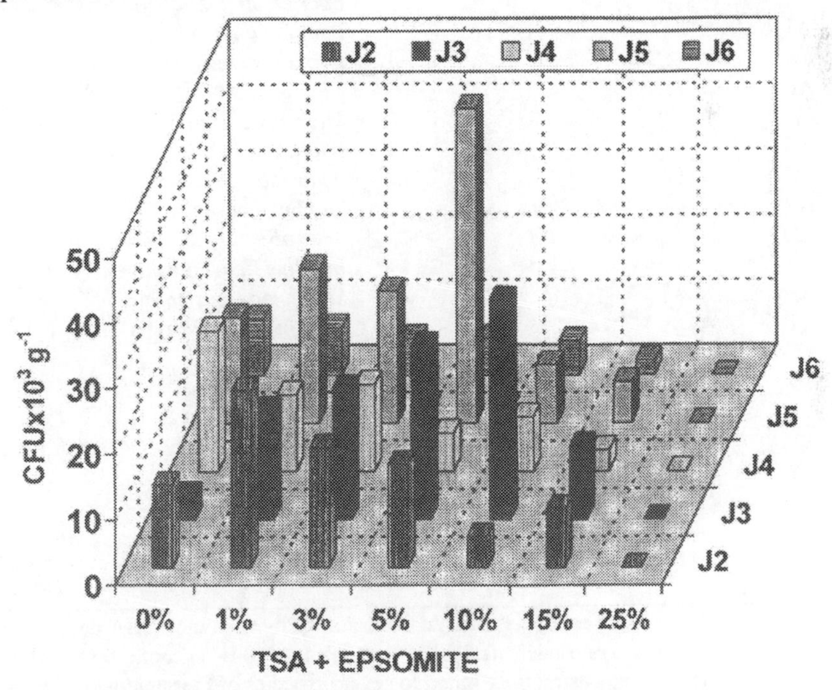

Figure 2. Colony forming units in samples inoculated in TSA + espomite

Although after 72 h of incubation bacteria did not appear at the highest salt concentrations (25%), two isolates were obtained after 15 days, indicating a poor and slow growth, which is normal for such a high salt concentration. Attempts to further cultivation were unsuccessful.

The isolates were characterized and identified using fatty acid analysis. Numerical analysis dendrograms were constructed for each sample and joined as shown in Table 1, which shows the grouping and identification of the isolates from efflorescences.

Table 1. Grouping and identification of 63 isolates from efflorescences samples

Cluster	J2	J3	J4	J5	J6	Identification with the TSBA library*
	Number of isolates					
AA		1			2	*Kocuria* sp.
AB		2				*Staphylococcus haemolyticus*
AC			3			No match
AD	2					*Bacillus* sp.
AE	1	1				*Bacillus pumilus* GC subgroup B
AF		1			1	*Bacillus freudenreichii*
AG	1	2	2		2	*Bacillus licheniformis*
AH					1	*Bacillus subtilis*
AI			1			*Bacillus amyloliquefaciens*
AJ			1			*Bacillus anthracis*
AK					1	*Bacillus cereus*
AL			1			No match
AM				1		*Micrococcus halobius*
AN		2			1	*Bacillus / Paenibacillus*
AO	1				1	*Arthrobacter* sp.
AP		2				*Staphylococcus* sp.
AQ		1	1			*Arthrobacter / Bacillus*
AR			1		1	*Bacillus licheniformis*
AS		2				*Staphylococcus cohnii*
AT			3			*Bacillus licheniformis*
AU		2				No match
AV	1					*Micrococcus* sp.
AW				2		*Paenibacillus* sp.
AX		2		1		*Bacillus sphaericus* GC subgroup III
AY			2			*Bacillus brevis*
AZ		1		1	1	No match
BA		1	1	2		*Bacillus* sp.
BB	1		1			*Bacillus megaterium*
Unclustered				1		

* Identification is given for each cluster, unless the similarity with the TSBA database is lower then 0.300. If clusters contain strains that are assigned to two genera, both names are reported. If clusters contains strains assigned to several species of the same genus, only the genus name is given.

The endospore-forming isolates were mainly *Bacillus* species. The genus *Bacillus* is the most abundant in normal and hypersaline soils, amounting for up to 67% and 19%, respectively (Rodriguez-Valera, 1988b), and in efflorescences from Romanian monasteries (Lazar and Dumitru, 1973). This genus is virtually absent in hypersaline waters. *Bacillus* encompasses the rod-shaped bacteria capable of aerobically forming endospores that are more resistant than vegetative cells to heat, drying, and other destructive agents. Use of fatty acid analysis resulted in the identification of 7 strains of *B. licheniformis*, 3 of *B. megaterium*, and 1 strain each of *B. anthracis*, *B. athrophaeus*, *B. marinus*, *B. pumilus*, *B. subtilis*, *B. cereus*, *B. amyloliquefaciens*, *B. sphaericus* and *B. brevis* (Table 1). The genera *Staphylococcus*, *Kocuria*, *Micrococcus*, *Paenibacillus* and *Arthrobacter* were also represented, with 3 strains of *S. cohnii* and 2 of *K. varians*. Thirty strains were not identified, which represented 48% of the isolates.

A general characteristic of hypersaline (aquatic) environments is that they are extreme environments and therefore species diversity is low, and some taxonomic groups are missing. However, a large microbial population was found in hypersaline soils with cell numbers ranging from 10^3-10^6 per gram of soil. Almost all the organisms isolated were halophilic eubacteria with a remarkable euryhaline response (Rodriguez-Valera, 1988b). Similar data were obtained for the efflorescences.

Bacteria are frequently exposed to stresses due to limitations and changes in nutrient availability, temperature, salinity, etc. and their persistence in the environment is to a great part determined by their capacity to endure these stresses. Osmotic adaptation of halophilic and halotolerant bacteria requires osmoadaptation strategies. While *Archaea* tolerate high cytoplasmic concentrations of KCl, the KCl-type strategy, the organic-osmolyte type is widespread among aerobic eubacterial halophiles (Galinski, 1993). Organic osmolytes responsible for osmotic balance and compatible with the cells' metabolism have been named compatible solutes, the most important being glycine betaine, ectoines, proline, *N*-acetylated diamino acids and *N*-derivatized glutamine amides. Most of these compounds have been found in bacilli which would explain the abundance of *Bacillus* species in efflorescences. Thus, the ability to synthesize ectoine is very common among aerobic heterotrophic eubacteria including Gram-positive cocci and *Bacillus* species; proline was originally considered the typical solute of halophilic *Bacillus* species, and a screening of halophilic/halotolerant bacilli and related species revealed that the majority of species produced ectoine, either alone or in combination with proline and/or acetylated diamino acids (Müller, 1991).

In the biosphere, bacteria can act as agents which disperse, fractionate or concentrate material. When functioning as dispersion agents, they can promote dissolution of insoluble minerals. As concentrating agents they can accumulate inorganic material via intra or extracellular deposition, and produce biomineralisation. This is the process through which organisms become involved in the formation of minerals. It is known that organisms form more than 60 biogenically different minerals.

It is well-known that bacteria are able to precipitate calcite and other minerals such as aragonite, struvite, bobierrite, apatite, etc. (Rivadeneyra et al., 1992). In this respect, the production of calcite crystals by soil bacteria has been considered a general phenomenon. Many bacteria, including *B. pumilus, B. subtilis, B. megaterium, B. cereus, B. bulgaricus, B. globigii, Pseudomonas aeruginosa, P. fluorescens, Serratia* spp., *Citrobacter* spp., *Azotobacter* spp., etc., formed calcite crystals on solid media containing sodium acetate (Boquet et al., 1973).

Rivadeneyra et al. (1993) found that 26 moderately halophilic strains of *Bacillus* isolated from a saline soil precipitated magnesium calcite, aragonite, monohydrocalcite and dolomite in varying proportions. The formation of calcium carbonate by some moderately halophilic bacteria was also observed by Del Moral et al. (1987). Optimum salt concentration for precipitation was 10%, precipitation decreasing at both 20 and 2.5% salt concentration.

Recently, Laiz et al. (1999) and Cañaveras et al. (1999) studied the formation of calcite, aragonite and hydromagnesite in caves. The presence of some of these minerals was highly suggestive of a microbially-mediated precipitation. To this end, crystal formation with all isolated bacteria from dripping waters and ceiling rock was tested. The number of isolates precipitating crystals ranged from 15 to 44% of all isolates, among which *B. cereus, B. sphaericus, Streptomyces rishiriensis, Chryseomonas luteola, Flavimonas oryzihabitans, Serratia liquefaciens, Xanthomonas maltophilia, Acinetobacter* sp. and *Amycolatopsis* sp. were identified. Crystal composition of the precipitate produced by cultures of *Acinetobacter* sp. resulted to be vaterite (85%) and calcite (15%). Although calcite was precipitated by different genera of bacteria, vaterite is highly unstable and is rarely found. Lowenstam (1981) only reported vaterite in *Rhodophyta*, in addition to a few animal taxa.

To what extent the above reported data can be extrapolated to nature? Lazar (1971) inoculated a dense bacterial suspension (3-4 x 10^{12} cells/ml) on undecayed frescoes zones and reproduced efflorescence formation. Although the studies have stressed the precipitation of crystals in culture media by aquatic and terrestrial bacteria (Laiz et al., 1999), this process needs to be proved *in vivo*, because the conditions used in the laboratory are far different from those that these bacteria find in their natural habitat. Furthermore, the sources of carbon and energy in culture media probably are different and exceed those available in nature. Although the ecology and physiology of bacteria in efflorescences is relatively unknown, the reported data indicate the presence of halotolerant bacteria, particularly *Bacillus* species, but more work is needed to ascertain the role of these bacteria in monuments. In addition, search for *Archaea* in deteriorated monuments should be carried out as there is an increasing evidence that these could be an appropriate environments for colonisation by and growth of these microorganisms.

ACKNOWLEDGEMENT

This work was supported by the European Commission, project ENV4-CT98-0705.

REFERENCES

Bassi, M. and Giacobini, C. 1973. Scanning electron microscopy: a new technique in the study of the microbiology of work of art. Int. Biodet. Bull. **9:** 57-68.

Boquet, E., Boronat, A. and Ramos-Cormenzana, A. 1973. Production of calcite (calcium carbonate) crystals by soil bacteria is a general phenomenon. Nature **246**: 527-528.

Cañaveras, J.C., Hoyos, M., Sanchez-Moral, S., Sanz-Rubio, E., Bedoya, J., Soler, V., Groth, I., Schumann, P., Laiz, L., Gonzalez, I. and Saiz-Jimenez, C. 1999. Microbial communities associated with hydromagnesite and nedle-fiber aragonite deposits in a karstic cave (Altamira, Northern Spain). Geomicrobiol. J. **16**: 9-25.

Charola, A.E. and Lewin, S.Z. 1979. Efflorescences on building stones-SEM in the characterization and elucidation of the mechanisms of formation. SEM/1979/I, SEM, Inc AMF O'Hare, IL 60666 p. 379-387

Del Moral, A., Roldan, E., Navarro, J., Monteoliva-Sanchez, M. and Ramos-Cormenzana, A. 1987. Formation of calcium carbonate crystals by moderately halophilic bacteria. Geomicrobiol. J. **5**: 79-87.

Galinski, E.A. 1993. Compatible solutes of halophilic eubacteria: molecular principles, water-solute interaction, stress protection. Experientia **49**: 487-496.

Gonzalez, I., Laiz, L., Hermosin, B., Caballero, B., Incerti, C. and Saiz-Jimenez, C. 1999. Bacteria isolated from rock art paintings: the case of Atlanterra shelter (south Spain) J. Microbiol. Meth. **36**: 123-127.

Incerti, C., Blanco-Varela, M.T., Puertas, F. and Saiz-Jimenez, C. 1997. Halotolerant and halophilic bacteria associated to efflorescences in Jerez cathedral. *In* F. Zezza (ed.), Origin, Mechanisms and Effects of Salts on Degradation of Monuments in Marine and Continental Environments, Protection and Conservation of the European Cultural Heritage Research Report nº 4 p. 225-232.

Javor, B. 1989. Hypersaline Environment: Microbiology and Biogeochemistry. Springer-Verlag, Berlin.

Laiz, L., Groth, I., Gonzalez, I. and Saiz-Jimenez, C. 1999. Microbiological study of the dripping waters in Altamira cave (Santillana del Mar, Spain). J. Microbiol. Meth. **36**: 129-138.

Lazar, I. 1971. Investigations on the presence and role of bacteria in deteriorated zones of Cozia monastery painting. Rev. Roum. Biol. Bot. **16**: 437-444.

Lazar, I. and Dumitru, L. 1973. Bacteria and their role in the deterioration of frescoes of the complex of monasteries from northern Moldavia. Rev. Roum. Biol. Bot. **18**: 191-197.

Lowenstam, H.A. 1981. Minerals formed by organisms. Science **211**: 1126-1131.

Mora, P., Mora, L. and Philippot, P. 1977. La Conservation des Peintures Murales. Editrice Compositori, Bologna.

Müller, E. 1991. Kompatible Solute und Prolingewinnung bei halophilen und halotoleranten Bacilli. Ph. D. Dissertation. Universität Bonn.

Rivadeneyra, M.A., Delgado, R., Delgado, G., del Moral, A., Ferrer, M.R. and Ramos-Cormenzana, A. 1993. Precipitation of carbonates by *Bacillus* sp. isolated from saline soils. Geomicrobiol. J. **11**: 175-184.

Rivadeneyra, M.A., Perez-Garcia, I. and Ramos-Cormenzana, A. 1992. Struvite precipitation by soil and fresh water bacteria. Curr. Microbiol. **24**: 343-347.

Rodriguez-Gordillo, J., Martin-Vivaldi Martinez, J.A. and Saiz-Jimenez, C. 1988. Estudio de los procesos de alteración, ensayos de envejecimiento acelerado y respuesta al tratamiento con diversos agentes preservantes, de los materiales pétreos de la cúpula de la Colegiata de Jerez de la Frontera (Cádiz). Congreso Geológico de España, Comunicaciones **2**: 341-343.

Rodriguez-Valera, F. 1988a. Halophilic Bacteria. CRC Press, Boca Raton, Florida.

Rodriguez-Valera, F. 1988b. Characteristics and microbial ecology of hypersaline environments. *In* Halophilic Bacteria, F. Rodriguez-Valera (ed.), CRC Press, Boca Raton, Florida p. 3-30.

Rodriguez-Valera, F., Ruiz-Berraquero, F. and Ramos-Cormenzana, A. 1981. Characteristics of the heterotrophic bacterial populations in hypersaline environments of different salt concentrations. Microb. Ecol. **7**: 235-243.

Rölleke, S., Muyzer, G., Wawer, C., Wanner, G. and Lubitz, W. 1996. Identification of bacteria in the biodegraded wall painting by denaturing gradient gel electrophoresis of PCR-amplified gene fragments coding for 16S rRNA. Appl. Environ. Microbiol. **62**: 2059-2065.

Rölleke, S., Witte, A., Wanner, G. and Lubitz, W. 1998. Medieval wall paintings-a habitat for *Archaea*: identification of *Archaea* by denaturing gradient gel electrophoresis (DGGE) of PCR-amplified gene fragments coding for 16S rRNA in a medieval wall painting. Int. Biodet. Biodeg. **41**: 85-92.

Saiz-Jimenez, C. 1982. Causas del deterioro de los murales de Daniel Vázquez Díaz, Monasterio de Santa María de la Rábida, Huelva. Mundo Científico **18**: 1007-1011.

Saiz-Jimenez, C. and Samson, R.A. 1981. Microorganisms and environmental pollution as deteriorating agents of the frescoes of the monastery of Santa Maria de la Rabida, Huelva, Spain. 6th Triennial Meeting ICOM, paper 81/15/5, 14 p.

Tiano, P., Bianchi, R., Gargani, G. and Vannucci, S. 1975. Research on the presence of sulphur-cycle bacteria in the stone of some historical buildings in Florence. Plant Soil **43**: 211-217.

Van der Marel, H.W. and Beutelspacher, H. 1976. Atlas of Infrared Spectroscopy of Clay Minerals and their Admixtures. Elsevier, Amsterdam.

Vreeland, R.H. and Hochstein, L.I. 1992. The Biology of Halophilic Bacteria. CRC Press, Boca Raton, Florida.

CHARACTERISATION OF ENDOLITHIC COMMUNITIES OF STONE MONUMENTS AND NATURAL OUTCROPS

Ornella Salvadori
Soprintendenza per i Beni Artistici e Storici di Venezia - Laboratorio Scientifico – Cannaregio 3553 – I-30131 Venezia, Italy

Key words: endoliths, cyanobacteria, fungi

Abstract: Different Italian monuments and natural outcrops have been investigated to check the presence of endolithic microorganisms. The colour of the examined surfaces varied from light grey to black. The results show the presence of cyanobacteria and fungi and suggest that endolithic communities are more widespread than previously thought. A first characterisation of their distribution patterns inside the stone is given.

1. INTRODUCTION

A particular aspect of biodeterioration of stone monuments is linked to endolithic organisms colonising the interior of rocks. Among them, endolithic lichens have been often recognised, described, and studied as they are easily distinguishable owing to the presence of the reproductive structures, the furrows, and sometimes the colour of the thalli (Tretiach, 1995; Pinna et al., 1998). On the contrary, the presence of endolithic microorganisms such as cyanobacteria, algae and fungi on monuments is only rarely reported. This is mainly due to the conventional techniques that are generally applied to study biodeterioration such as microbiological cultures, and observation by optical microscope of microorganisms alone, not in relation with the stone. Also the molecular techniques, more and more applied in the field of conservation, could be very useful to detect and identify different microbial strains but they do not help in the evaluation of damage induced on the substrate.

On the contrary, many studies were performed on endolithic microflora mainly in marine environment or in extreme environments like deserts and Antarctica. It is well known that marine boring algae and fungi colonise a variety of carbonate substrate (calcareous rocks, shells, skeletal fragments, sand grains and oolites) and form dark-coloured bands along the coastal limestone (Golubic, 1969, 1976; Kohlmeyer, 1969). Boring algae (or perforating algae) contribute to the destruction of carbonate coasts, an active penetration seems limited to more soluble rocks such as carbonatic rocks (Golubic and Schneider, 1972; Golubic and Le Campion-Alsumard, 1973; Golubic et al., 1975; Schneider, 1976).

Sub-aerial endolithic algae were much less studied than marine ones (Bachmann, 1915; Pia, 1937; Degelius, 1962; Danin and Garty, 1983). Friedmann (1971) and Friedmann et al. (1967, 1988) described cryptoendolithic cyanobacteria developing in extremely dry habitats like the Negev Desert (Israel) and the Death Valley (California). The organisms develop between the sandstone crystals forming extended layers a few mm deep (from 0.1 to 2.5 mm) under the rock surface. The species with this kind of growth seem limited to a single genus, *Gloeocapsa*.

The occurrence of phototrophic microorganisms on stone monuments has been recognised very frequently and many cyanobacteria, green algae and diatoms were identified (Ortega-Calvo et al., 1991, 1993). Nevertheless, only few authors reported the presence of endolithic photosynthetic microorganisms on monuments: *Hyella fontana* living in the marble of some outside sculptures of the Ostia Antica Museum (Italy) (Giaccone et al., 1976) and in the Lindaraja Fountain, Granada (Spain) (Bolivar and Sanchez-Castillo, 1997), a cryptoendolithic microbial community in a limestone of the Basilica of Tongerem (Belgium), *Stichocococcus* sp. and *Chroococcidiopsis* sp. (Saiz-Jimenez et al., 1990); *Phormidium* sp. under black sulfated crusts on limestone of the cathedral of Seville (Saiz-Jimenez, 1995).

Golubic et al. (1981) proposed a terminology, now generally accepted, combining topical and functional criteria describing the lithobiontic ecological niche. They divided endoliths into three groups: chasmoendoliths (colonising fissures and cracks), cryptoendoliths (colonising structural cavities within porous rocks) and euendoliths or rock boring organisms (penetrating actively into the interior of rocks and forming tunnels).

Endolithic microrganisms are generally described as slow-growing, stress-tolerant organisms as they can survive in inhospitable habitats with high temperatures, UV-radiations and sometimes drought. As regards artworks, they are spread not only in particular environments where the humidity or the water availability are high (fountains, artworks in contact with the soil, and often surrounded by a lot of vegetation) but also in dryer environments such as vertical surfaces of buildings.

This paper is primarily concerned with the descrption of the most useful methods for studying endolithic communities inhabiting stone and for characterising their pattern of distribution. Due to the intimate association between endoliths and the substrate, different techniques are necessary to study them: endolithic microorganisms can be observed after dissolution of rock, but this causes the dispersion of filaments or cells and a loss of information about their spatial distribution so it is often difficult to understand the relationship within the microbial communities. The association of petrographic techniques as well as SEM observations of resin casts after a partial removal of the substrate permit to observe the interrelations between microorganisms and stone.

2. RESULTS

Endoliths produce alterations of the stone surface very similar in appearance to others of abiotic genesis and visual examination often reveals no evidence of biological colonisation. This may be one of the reasons of the few researches regarding these microorganisms. They induce (frequently in parts flushed by rainwater) a change in the colour of the stone that becomes darker, generally from light grey to black, and many meters of the exposed surfaces or the entire artwork (e.g. tombstone) can be interested by this kind of phenomenon. A superficial patina can be rarely recognised and it is generally impossible to take samples without scraping or removing small slabs of stone. Sometimes "the problem" appears during or after cleaning operations when the restorers notice that the stone seems clean but that wetting or the application of resins cause a colour change.

Different specimens were examined, from Italian monuments and natural outcrops: Nettuno fountain (Trento), tombstones of the Old Jewish Cemetery (Venice), some internal facades of Ducal Palace (Venice), and marbles from Carrara's quarries (Tab.1). The examination of samples demonstrated that in all of them black crusts or other inorganic deposits are present, even if it has been previously stated that they can support the colonisation of phototrophic microorganisms and fungi (Ortega-Calvo et al., 1991, 1995).

Small specimens were first fixed in 4% formaldehyde solution and then put in contact with a dilute solution of hydrochloric acid to remove calcium carbonate. After stone dissolution, the remaining part was observed by light transmission microscopy. Reflected light microscopy was carried out on polished cross sections, freshly prepared and stained with Periodic Acid Schiff (PAS) according to Whitlack and Johnson (1974). Thin sections were studied with a polarising microscope. For scanning electron microscopy samples were prepared according to the methods used for studying marine endolithic algae: samples were fixed overnight with 2.5% glutaraldehyde in 0.01 M phosphate buffer, serially dehydrated in ethanol, included in epoxy resin and then treated with Perenyi solution (Le Campion-Alsumard, 1976, 1979). The observation of vertically fractured samples with a scanning electron microscope permits to have an idea about the penetration of microorganisms, but only the preparation of resin casts allows to appreciate the exact morphology and diffusion of endolithic communities (Fig 2, 3, 4).

Table 1. Localities, rock type and colour of surface of examined samples.

Samples	Description	Substrate	Surface appearance
EN1	Carrara's quarry	Marble	Black
EN2	Carrara's quarry	Marble	Grey
EN3	Tombstone, Old Jewish Cemetery, Venice	Istrian stone	Light grey
EN4	Carrara's quarry	Marble	Dark grey
EN5	Tombstone, Old Jewish Cemetery, Venice	Istrian stone	Grey
EN6	Nettuno fountain, Trento	Limestone	Light grey
EN7	Ducal Palace, internal facade, Venice	Verona Red stone	Grey
EN8	Ducal Palace, internal facade, Venice	Istrian stone	Light grey
EN9	Tombstone, Old Jewish Cemetery, Venice	Istrian stone	Grey
EN10	Tombstone, Old Jewish Cemetery, Venice	Istrian stone	Grey

2.1 Marble from Carrara's quarries

Sample EN1 – The marble shows a high decohesion and calcite crystals are surrounded by black materials giving a black appearance to the surface. Polished cross sections show colonisation by cyanobacteria penetrating to a depth of 110-150 μm. Algal cells diversely pigmented run along intergranular spaces until 1.5 mm. Only in some superficial limited zones black fungi are present. After PAS staining, a thick net of fungi surrounded by a considerable amount of extracellular polymeric substances (EPS) appears around crystals to a depth of 750 μm. After dissolution, a lot of gelatinous materials is appreciable among algae and fungi. Only the cyanobacteria colonising the upper surface are euendolithic (rock boring) (Fig. 2), other algae and fungi have to be considered chasmolithic as developing only in the intergranular spaces.

Sample EN2 – The marble surface is grey and cribriform and shows a diffuse pitting with pits of different dimensions appearing lighter than the not-pitted surfaces. Little black clusters of dark pigmented fungi are present on the surface. Polished cross sections show the development of endolithic cyanobacteria to a depth of 110-150 μm. Dark-pigmented hyphae are very scant and penetrate to a maximum of 380 μm.

Sample EN4 – The marble surface is dark grey and cribriform; in fractured samples 2 mm from the rock surface a dark brown zone horizontally extends and is 0.3-2 mm thick (Fig. 1). Euendolithic cyanobacteria thickly colonise marble and penetrate down to 120-250 μm, other algae are recognisable to a depth of 1.3 mm.

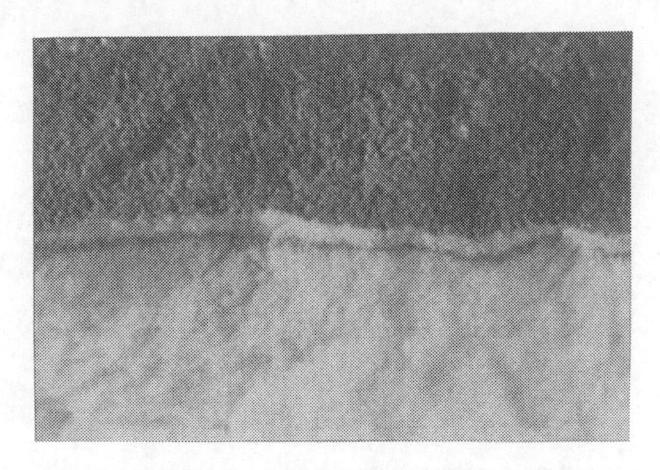

Figure 1. Fractured Carrara marble showing a dark grey surface and a brown zone developing 2 mm from the surface.

The brown layer parallel to the marble surface is due to the presence of filamentous microorganisms and to a lot of extracellular substances, both coloured in brown. PAS staining demonstrated a deep development of filamentous colourless fungi among marble grains to a depth of more than 5 mm.

2.2 Nettuno fountain (Trento)

The restoration of Nettuno fountain (1767-1769) started at the beginning of 1991, and, after the cleaning operation, the restorers noticed that large areas of limestone became greenish on wetting even if the surfaces appeared well cleaned. The analyses revealed the presence of filamentous cyanobacteria up to 400 μm in some zones to a maximum of 800 μm in others. Some filaments were also recorded at a depth of 1.9 mm. The filaments open to the limestone surface but they did not have an appreciable development on stone.

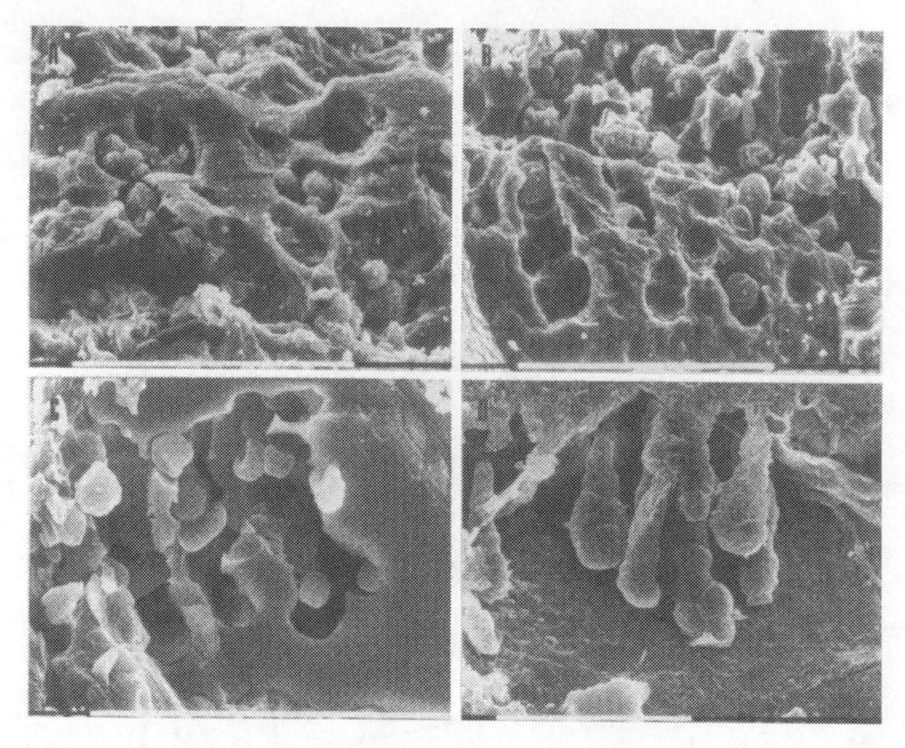

Figure 2. Scanning electron micrographs of endolithic cyanobacteria boring Carrara marble. (A) openings at the surface of the bored tunnels; (B) and (C) vertically fractured sample with perforation induced by cyanobacteria; (D) resin casts of photosynthetic clusters boring the upper part of marble. Scale bars indicate 0.1 mm.

2.3 Ducal Palace, internal facades of the courtyard (Venice)

More or less extended areas of the internal facades are grey or light grey coloured, but this is not due to a superficial deposit. Moreover these zones are clearly distinguishable from black crusts or other kind of inorganic deposits, also considering their localisation and water availability. Polished cross sections showed the presence of endolithic cyanobacteria to a depth of 150 μm. After PAS staining the presence of a loose web of hyphae especially in Verona Red stone was detected, whereas in Istrian stone only a few fungal filaments were occasionally present.

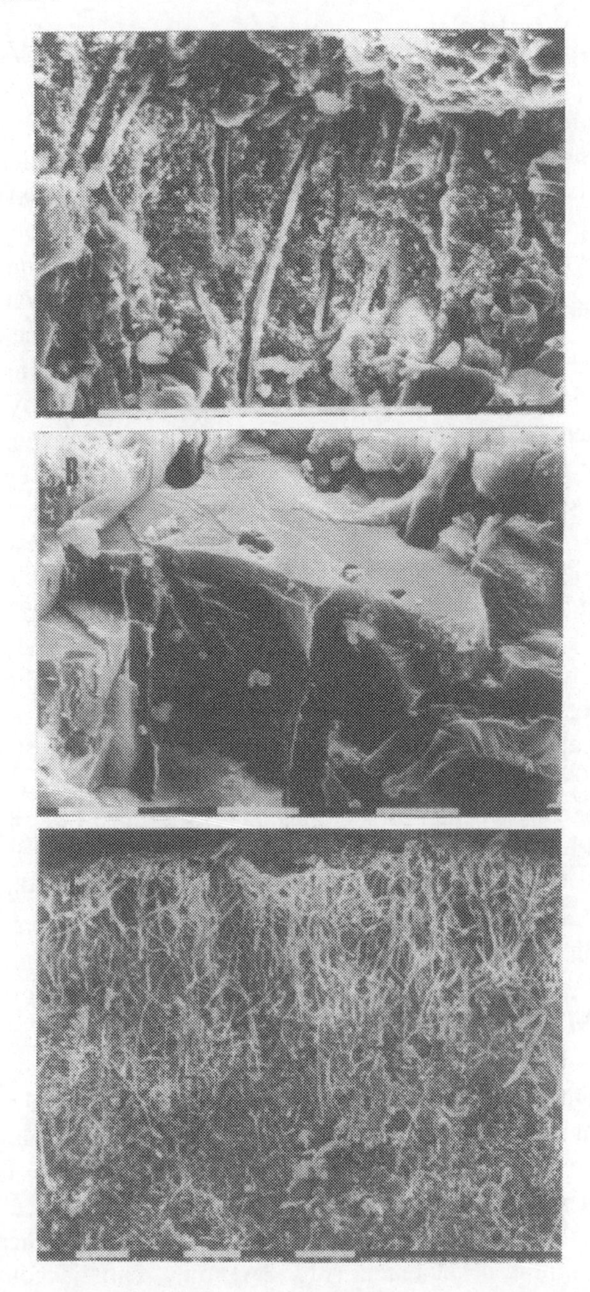

Figure 3. SEM photographs of filamentous cyanobacteria boring Nettuno fountain. (A) and (B) portions of the vertically fractured limestone showing the bored tunnels; (C) resin cast of the algal distribution pattern within the stone. Scale bars indicate 0.1 mm in (A) and (C), and 10 μm in (B).

2.4 Tombstones of the old Jewish Cemetery (Venice)

All the examined tombstones from the Old Jewish Cemetery (Venice) are colonised by dark pigmented fungi, generally described as dematiaceous or black fungi. They show an epilithic development, in some cases reaching a thickness of 60-100 μm, and form a compact layer of multiseriated pseudoparenchymatous cells remaining entire also after stone dissolution. From this layer endolithic hyphae penetrate into the substrate; the fungal pigmentation is very intense in the hyphae on the stone surface and generally decreases in the deepest ones appearing colourless. Occasionally, only few cells of epilithic cyanobacteria and algae were observed by transmission microscopy after rock dissolution.

The role of fungi in rock weathering was stated by many authors and recently the involvement of microcolonial fungi or black dematiaceous fungi in stone weathering was underlined. These fungi grow very slowly and have a dark pigmentation useful against UV irradiation (Soukharjevski et al., 1994; Saiz-Jimenez et al., 1995). According to Gorbushina et al. (1997) they tend to penetrate into materials or to create fissures, cavities and biopits. The importance of fungi in the production of black-grey patina and their ability of penetrating deeply in the rock was previously stressed (Urzì et al., 1993, 1994; Diakumaku et al., 1994; Wollenzein et al., 1995; Krumbein and Diakumaku, 1996; Urzì and Realini, 1998) but an intense growth like that registered on and in Venetian tombstones has never been described before. In the case of Old Jewish Cemetery, the location of tombstones on and often partially buried in the soil, and the influence of surrounding thick vegetation surely ensure an ample availability of nutrients and water supply thus stimulating such a growth.

2.5 SEM observations

SEM examination of vertically fractured samples permitted to observe the development of microorganisms on the surface of the samples and to appreciate some details of their penetration into the substrate. In the case of the samples of Carrara marble, Nettuno fountain and Ducal Palace, the photosynthetic clusters of cyanobacteria open to external surface but they do not have any epilithic development (Fig. 2A); only dematiaceous fungi grow on and within tombstones.

In marble samples, it is particularly evident that euendolithic cyanobacteria and fungi can easily penetrate the calcite crystals. The size of the bored tunnels depends on the size of the endolithic filaments: 15-25 μm for cyanobacteria developing on Carrara marble, 3-5 μm for filamentous cyanobacteria of Nettuno fountain and 5-8 μm for dark-pigmented fungi (Fig. 2, 3, 4).

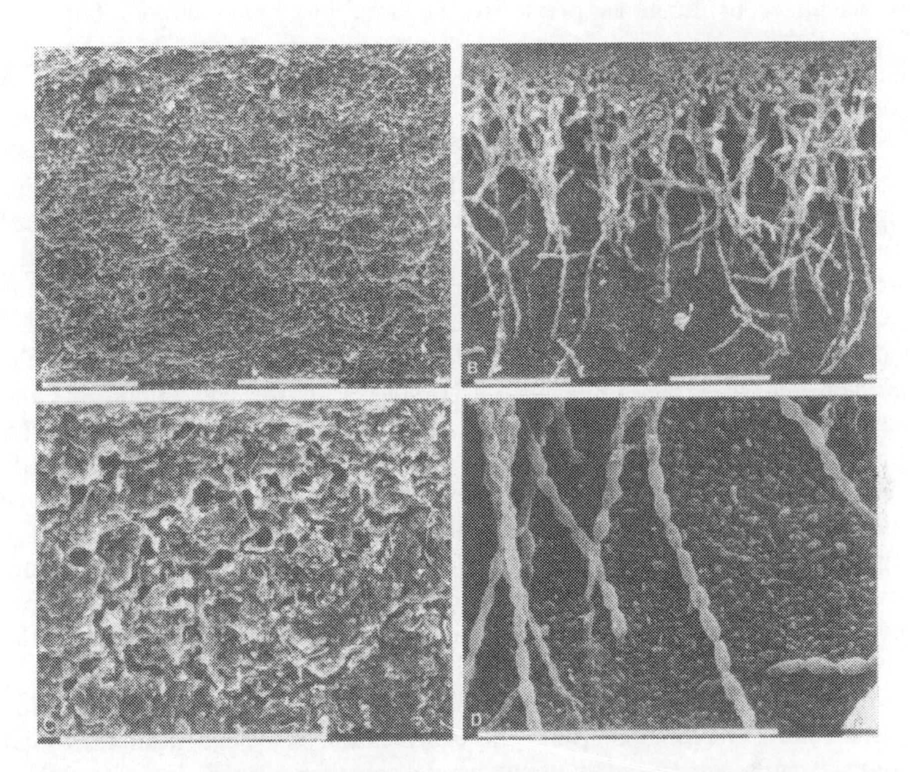

Figure 4. Scanning electron micrographs of tombstones colonised by dematiaceous fungi. With vertically fractured samples (A) and (C) only some holes and tunnels bored by fungi are appreciable; resin cast shows clearly the fungal distribution pattern (B); enlarged detail of endolithic hyphae (D). Scale bars indicate 0.1 mm.

Resin casts permit to appreciate the arrangement of endolithic communities inside the stone and some morphological characteristics. The different level of information obtained with vertically fractured samples or resin casts is evident comparing scanning electron micrographs at the same magnification (Fig. 4A and 4B).

3. CONCLUSIONS

Endolithic microorganisms have been found in all examined samples coming from different environments and substrates, in which their presence was hypothesised. This suggests that endolithic communities are more widespread than previously thought and that their role in stone biodeterioration has been underestimated. The grey-black colour appearance of stone surfaces, when biologically induced, is not only due to the presence of dematiaceous fungi as previously stressed by many authors but very frequently it is due to endolithic phototrophic microorganisms (cyanobacteria and algae). As in in the case of fungi, dark pigments protect algal cells against UV-irradiation and other stress factors.

It is interesting to note that free endolithic cyanobacteria on Carrara marble, Istrian stone and Red Verona limestone reach on the average the same depth of photobionts of endolithic lichens (Pinna et al., 1998). Filamentous cyanobacteria colonising the Nettuno fountain penetrate more into the substrate, as frequently described for marine phototrophic endoliths.

Notwithstanding some studies, many aspects of the physiology of endolithic microorganisms still await further investigation. For instance, the mechanism of penetration into the rock is not yet known: endolithic fungi and lichens seem to produce negligible amounts of organic acids compared with epilithic ones, and the physiology of endolithic cyanobacteria or algae is little known. Many dark pigmented fungi isolated from stone seem not to produce acids (Krumbein and Diakumaku, 1996) and for this reason these authors suggest that the physical attack could be the most important factor in the penetration mechanism. Further research will be necessary to clarify how endoliths (cyanobacteria, algae, fungi and lichens) penetrate into the substrate; probably, the mechanism is the same.

A first characterisation of some different distribution patterns of endolithic cyanobacteria and fungi is given. The identification of the endolithic microbial strains is in progress and it could not be very easy as their taxonomy is unclear. It is necessary to investigate methodically all the suspected presence of endolithic communities on monuments to better know and characterise them. In fact, too many colonised stones are still considered to be due to deposits of atmospheric pollutants. In addition, this study confirms the usefulness of applying different techniques to recognise and characterise all the microorganisms present on and in the stone.

A better knowledge of endolithic microorganisms is important either for an accurate evaluation of the damage caused by them or for a correct planning of restoration. Therefore the rate of penetration and the intensity of endolithic colonisation are important data to be considered before choosing any kind of intervention.

REFERENCES

Anagnostidis, K., A. Economou-Amilli and M. Roussomoustakaki. 1983. Epilithic and chasmolithic microflora (Cyanophyta, Bacillariophyta) from marbles of the Partenon (Acropolis-Athens, Greece). Nova Edwigia **38:** 227-287.

Ascaso, C., J. Wierzchos and R. Castello. 1998. Study of the biogenic weathering of calcareous litharenite stones caused by lichen and endolithic microorganisms. Internatl. Biodet. Biodeg. **42:** 29-38.

Bachmann, E. 1915. Kalklösende algen. Ber. Deutsch. Bot. Ges. **31:** 3-12.

Bolivar, F.C., and P.M. Sanchez-Castillo. 1997. Biomineralization processes in the fountains of the Alhambra, Granada, Spain. Internatl. Biodet. Biodeg. **40:** 205-215.

Danin, A. and J. Garty. 1983. Distribution of cyanobacteria and lichens on hillsides of the Negev Highlands and their impact on biogenic weathering. Z. Geomorph. N.F. **27:** 423-444.

De Leo, F., G. Criseo and C. Urzì. 1996. Impact of surrounding vegetation and soil on the colonization of marble statues by dematiaceous fungi. *In* Proc. of the 8th Int. Cong. On Deterioration and Conservation of Stone, J. Riederer (ed.), Berlin p. 625-630.

Degelius, G. 1962. Über verwitterung von kalk- und dolomitgestein durch algen und flechten. *In* Chemie im dienst der archäologie bautechnik denkmalpfleg, J.A. Hedvall (ed.), Lund

Diakumaku E., P. Ausset, K. Sterflinger, U. Wollenzien, W.E. Krumbein and R.A. Lefèvre. 1994. On the problem of rock blackening by fly-ash, fungal and other biogenic particles, and their detection in Mediterranean marbles and monuments. *In* III International Symposium on the Conservation of Monuments in the Mediterranean Basin, V. Fassina, H. Ott, and F. Zezza (eds.), Venice p. 305-310.

Friedmann, E.I. 1971. Light and scanning electron microscopy of the endolithic desert habitat. Phycologia **10:** 411-428.

Friedmann, E.I., M. Hua and R. Ocampo-Friedmann. 1988. Cryptoendolithic lichen and cyanobacteral communities of the Ross Desert, Antartica. Polarforschung **58:** 251-259.

Friedmann, E.I., Y. Lipkin and R. Ocampo-Paus. 1967. Desert algae of the Negev. Phycologia **6:** 185-196.

Giaccone, G., M.L. Veloccia Rinaldi and C. Giacobini. 1976. Forme biologiche delle alghe esistenti sulle sculture all'aperto. *In* R. The Conservation of Stone I, Rossi Manaresi (ed.), Centro per la conservazione delle sculture all'aperto, Bologna p. 245-256.

Golubic, S. 1969. Distribution, taxonomy and boring patterns of marine endolithic algae. Am. Zool. **9:** 747-751.

Golubic, S. 1976. The relationship between blue-green algae and carbonate deposits. *In* The biology of blue-green algae, N. Carr, and B. Whitton (eds.), Blackwell Sci. Publ., London p. 434-472.

Golubic, S., G. Brent and T. Le Campion-Alsumard. 1970. Scanning electron microscopy of endolithic algae and fungi using a multipurpose casting – embedding technique. Lethaia **3:** 203-209.

Golubic, S., E. Friedmann and J. Schneider. 1981. The lithobiontic ecological niche, with special reference to microorganisms. J. Sedim. Petr. **51:** 475-478.

Golubic, S. and T. Le Campion-Alsumard. 1973. Boring behaviour of marine blue-green algae *Mastigocoleus testarum* Lagerheim and *Kyrtuthrix dalmatica* Ercegovic as a taxonomic character. Schweiz. Z. Hydrol. **35:** 157-161.

Golubic, S., R.D. Perkins and R. Lukas. 1975. Boring microorganisms and microborings in carbonate substrate. *In* Study of trace fossils, R. Frey (ed.), Springer-Verlag, New York p. 229-259.

Golubic, S. and J. Schneider. 1972. Relationship between cabonate substrate and boring patterns of marine microorganisms. *In* Geol. Soc. Am., Ann. Meet. Abstr. Progr. p. 518.

Gorbushina, A.A., W.E. Krumbein and D.Yu. Vlasov. 1997. The fungal microcosm of Mediterranean monuments and sites-past, present and future. *In* Proc. 4th Int. Symp. on the Conservation of Monuments in the Mediterranean, A. Moropoulou, F. Zezza, E. Kollias, and I. Papachristodoulou (eds.), Rhodes, vol.4 p. 261-270.

Gutiérrez, A., M.J. Martinez, G. Almendros, F.J. Gonzalez-Vila and A.T. Martinez. 1995. Hyphal-sheath polysaccharides in fungal deterioration. The Science of the Total Environment **167:** 315-328.

Hirsch, P., F.E.W. Eckhardt and R.J Palmer, Jr. 1995. Fungi active in weathering of rock and stone monuments. Canadian J. Botany **73:** 1384-1390.

Hoffman L. 1989. Algae of terrestrial habitats. Bot. Rev. **55:** 77-105.

Kobluk, D.R. and M.J. Risk. 1977. Rate and nature of infestation of a carbonate substratum by a boring alga. J. Exp. Mar. Biol. Ecol. **27:** 107-115.

Kohlmeyer, J. 1969. The role of marine fungi in the penetration of calcareous substances. Am. Zoologist **9:** 741-746.

Krumbein, W.E. and E. Diakumaku. 1996. The role of fungi in the deterioration of stone. *In* Interactive physical weathering and bioreceptivity study on building stones, monitored by Computerized X-Ray Tomography (CT) as a potential non-destructive research tool. Protection and conservation of the European cultural heritage, Research Report n.2 p. 140-170.

Le Campion-Alsumard,T. 1976. Etude preliminaire sur l'ecologie et l'ultrastructure d'une Cyanophycee Chroococcale endolithe. J. Microscopie Biol. Cell. **26:** 53-60.

Le Campion-Alsumard,T. 1979. Les Cyanophycees endolithes marines. Systematique, ultrastructure, ecologie et biodestruction. Oceanologica Acta **2:** 143-156.

Ortega-Calvo, J.J., X. Arino, M. Hernandez-Marine and C. Saiz-Jimenez. 1995. Factors affecting the weathering and colonization of monuments by phototrophic microorganisms. The Science of the Total Environment **167:** 329-341.

Ortega-Calvo, J.J., M. Hernandez-Marine and C. Saiz-Jimenez. 1991. Biodeterioration of building materials by cyanobacteria and algae. Internal. Biodet. **28:** 167-187.

Ortega-Calvo, J.J., M. Hernandez-Marine and C. Saiz-Jimenez. 1993. Cyanobacteria and algae on historic buildings and monuments. *In* Recent Advances in Biodeterioration and Biodegradation, K.L. Garg, N. Garg, and K.G. Mukerji (eds.) Naya Prokash, Calcutta p. 173-203.

Pia, J. 1937. Die kalklõsende Thallophyten. Arch. Hydrobiol. **31:** 264-328.

Pinna, D., O. Salvadori and M. Tretiach. 1998. An anatomical investigation of calcicolous endolithic lichens from the Trieste karst (NE Italy). Plant Biosystems **132:** 183-195.

Saiz-Jimenez, C. 1995. Deposition of anthropogenic compounds on monuments and their effect on airborne microorganisms. Aerobiologia **11:** 161-175.

Saiz-Jimenez, C., X. Arino and J.J. Ortega Calvo. 1996. Mechanisms of stone deterioration by photosynthesis-based epilithic biofilms. *In* Interactive physical weathering and bioreceptivity study on building stones, monitored by Computerized X-Ray Tomography (CT) as a potential non-destructive research tool. Protection and Conservation of the European Cultural Heritage, Research Report n. 2 p. 25-62.

Saiz-Jimenez, C., J. Garcia-Rowe, M.A. Garcia Del Cura, J.J. Ortega Calvo, E. Roekens and R. Van Grieken. 1990. Endolithic cyanobacteria in Maastricht limestone. The Science of the Total Environment **94:** 209-220.

Saiz-Jimenez, C., J.J. Ortega Calvo and J.W. de Leeuw. 1995. The chemical structure of fungal melanins and their possible contribution to black stains in stone monuments. The Science of the Total Environment **167:** 305-314.

Schneider, J. 1976. Carbonate construction and decomposition by epilithic end endolithic microorganisms in salt- and freshwater. *In* The biology of blue-green algae, N. Carr and B. Whitton (eds.), Blackwell, London p. 248-260.

Slavoshevskaya L., O. Smolyanitskaya, V. Mozgovoy and S. Petrova. 1997. Mycological investigation of deteriorated ancient Greece and Rome marble monuments from the collection of the Hermitage Museum. *In* Proc. 4th Int. Symp. on the Conservation of Monuments in the Mediterranean, A. Moropoulou, F. Zezza, E. Kollias, and I. Papachristodoulou (eds.), Rhodes, vol.4 p. 437-451.

Soukharjevski, S., A.A. Gorbushina, W.E. Krumbein and L. Panina. 1994. Recognition and identification of marbles and marble infecting black fungi using EPR-techniques. *In* III International Symposium on the Conservation of Monuments in the Mediterranean Basin, V. Fassina, H. Ott, and F. Zezza (eds.) p. 335-341.

Staley, J.T., F. Palmer and J.B. Adams. 1982. Microcolonial fungi: common inhabitants on desert rocks? Science **215:** 1093-1095.

Tiano, P., P. Accolla and L. Tomaselli. 1995. Phototrophic biodeteriogens on lithoid surfaces: an ecological study. Microb. Ecol. **29:** 299-309.

Tretiach, M. 1995. Ecophysiology of calcicolous endolithic lichens: progress and problems. Giorn. Bot. Ital. **129:** 159-184.

Urzì, C., G. Criseo, W.E. Krumbein, U. Wollenzien and A.A. Gorbushina. 1993. Are colour changes of rocks caused by climate, pollution, biological growth, or by interactions of the three? *In* Conservation of Stone and Other Materials, M.J. Thiel (ed.), E. & F.N. Spon, London p. 279-286.

Urzì, C., W.E. Krumbein, N. Lyalikova, J. Petushkova, U. Wollenzien and M. Zagari. 1994. Microbiological investigation of marbles exposed to natural and anthropogenic influences in northern and southern climates. *In* III International Symposium on the Conservation of Monuments in the Mediterranean Basin, V. Fassina, H. Ott, and F. Zezza (eds.) p. 297-304.

Urzì, C. and M. Realini. 1998. Colour changes of Noto's calcareous sandstone as related to its colonisation by microorganisms. Internatl. Biodet. Biodeg. **40:** 45-54.

Whitlack R.B. and R.G. Johnson. 1974. Methods for staining organic matter in marine sediments. J. Sedim. Petr. **44:** 1310-1312.

Wollenzien, U., G.S. de Hoog, W.E. Krumbein and C. Urzì. 1995. On the isolation of microcolonial fungi occurring on and in marble and other calcareous rocks. The Science of the Total Environment **167:** 287-294.

PART 2

BIOSUSCEPTIBILITY OF ORGANIC AND INORGANIC CONSTITUENTS

PATINA

Physical and Chemical Interactions of Sub-aerial Biofilms with Objects of Art

Thomas Dornieden, Anna A. Gorbushina and Wolfgang E. Krumbein
Geomicrobiology, ICBM, Carl von Ossietzky Universitaet Oldenburg, P.O.B. 2503, D-26111 Oldenburg, Germany

Key words: biofilms, sub-aerial, chemical biodeterioration, patina, physical biodeterioration, poikilotrophic microorganisms

Abstract: In this contribution on the interaction of microorganisms with their substrates special attention is given to the development of microbial biofilms at and below the surface of monuments and objects of art. These biofilms can be very detrimental and thus quickly destroy the substrate. However, under certain conditions, the microflora contributes to the formation of a stabilised skin or layer. Slow growing poikilotrophic microorganisms are important factors in this process. This layer, which is often pigmented differently than the substrate, is called patina. Patina is regarded as the sum of all surface changes of an object of art. The interactions of the biofilms with the substrates are considered under physical, chemical and aesthetic aspects. Physical or mechanical attacks and alterations are often more important than chemical ones. Patina or biopatina by itself represents a considerable change of the original substrate. However, it may also be regarded as an hiatus period for the monument or object in question. The physical-chemical changes leading to a patina and sometimes to a subsequent breakdown of the upper layers of the material may sometimes reach almost geological time-scales.

1. INTRODUCTION AND DEFINITIONS

The interaction of organisms with their substrates is multiple and depends on many factors innate to the substrate, to the organisms themselves, and to the surrounding medium. Physical and chemical interactions embrace many means. The reception of the spoken word is a physically induced chemical process ending in a biological phenomenon while (1) the craquélé on ancient glass and (2) the pitting and disaggregation of ancient marbles is a chemically induced physical/mechanical process impossible to explain without considering the involvement of biological phenomena (Mellor, 1922; Krumbein, 1966; Dornieden et al., 1997). Objects of art consist of different materials, which serve in several respects as substrates (Brachert, 1995).

The term substrate has at least two meanings: (1) the space or surface in or on which an individual organism is settling and living and (2) the source of energy, electrons and elements or compounds needed as a substrate for growth and maintaining metabolic activity of the organisms living on the physical substrate. Microorganisms thus may use objects of art in both ways, either as a site of attachment and dwelling or as nutrient and energy source.

Sub-aerial biofilms are microbial communities exposed mainly to the atmosphere and less frequently to running or standing water. Sub-aquatic biofilms can be regarded as a semi-solid status of 99% water, solidified by a certain number of different microorganisms through extracellular polymeric substances (EPS). Sub-aerial biofilms in contrast are characterised as more than 99% of organic matter of microorganisms and their EPS maintaining metabolic potential at a minimum of available water and water activity (Gorbushina et al., 1999; Gorbushina and Krumbein, 1999a; Gorbushina and Krumbein, in press a, b). Only periodically or episodically such biofilms will gain full metabolic activity.

Such environments have been termed poikilotroph environments. Gorbushina and Krumbein (1999b) found by mycological analysis of 10 ancient Greek and Roman monument sites that sub-aerial biofilms on excavated and re-exposed architectural and sculptural objects are mainly composed of microorganisms representing the poikilotroph life style. Staley et al. (1982) were apparently the first to hint to this interesting group of microorganisms active in the colonisation of bare rock surfaces. Poikilotrophs are organisms permanently present in the environment. They grow slowly, but they survive under all circumstances and conditions present in any environment. This way in habitats with extremely changing environmental conditions through time they will slowly out-compete all other organisms fit for more favourable and mild environmental conditions. On the other hand, they also exist in oligotrophic and eutrophic conditions, a fact already indicated by Gromov (1963). Under these conditions they are, however, somehow hidden among the mass of more specialised organisms and are barely detected or isolated from such environments. Further the poikilotroph organism has a high potential of exerting pressure on the substrate in this harsh environment by shaping it mechanically and chemically through (1) growth pressure, (2) mechanical impacts through the cell wall and physical characteristics as e.g. pigmentation and other ways of making it more rigid, (3) extracellular polymeric compounds and metabolites, (4) very high morphological and (5) extremely high spatial flexibility.

Conservators and restorers try to prevent physical and chemical damage of the objects by keeping (1) the amount of water (humidity), and (2) organic and inorganic substances which can support growth and other conditions suitable for life processes at the possible minimum in view of the multiple detrimental interactions between microorganisms and the materials. The environment that conservators regard as optimal for their objects thus serves to select for poikilotrophic representatives of the microorganisms settling on and in objects of art. However, in many cases water activity and external and internal sources of organic and other nutrients are either unavoidable or occur for certain periods of time. Periods allowing increased activity of poikilotrophic microbiota are the true danger for objects of art. The poikilotroph will drastically change the physical substrate and partially use it as a nutritional substrate. This leads to the destruction of the substrate.

However, the interaction of microbiota persisting for long periods of time with the objects of art without detrimental action can also help to protect our cultural heritage. Organisms, which tend to keep themselves alive for hundreds if not thousands of years, will not always and not immediately be a damaging factor. One can even talk about growth on a geological time scale, which overlaps with the cultural heritage time scales at least in the case of pre-historic objects of art (Sun and Friedmann, 1999).

Mellor (1922) was only mentioning the detrimental effects of the biofilm, while Krumbein (1966) hinted to the protective action of pigmented biofilms changing the reactivity of the surface of cultural heritage objects. These changes were later followed up intensively by Krumbein (1969), Gorbushina et al. (1993), Diakumaku (1995) and Diakumaku et al. (1995). A biogenic and often biopigment stabilised patina is always a sign of an hiatus in the decay process and can be regarded as the best helper to the conservator/restorer as long as she/he accepts the almost mystical changes of colour and hue so characteristic for ancient objects.

Thus biofilms and their physical and chemical interactions with objects of art have to be analysed very carefully for they may be extremely detrimental and protective at the same time. The equilibrium between the material of which an object of art is made, the environment and the biota establishing themselves is a delicate one. However, we have witnessed cases in which a material doomed to physical and chemical decay was protected by the biological envelope developing for hundreds of years. Krumbein (1966, 1993, 1998) has summarised the potential interactions between (1) the materials, (2) organismic and mainly microbial communities settling on and within, and (3) the environmental factors influencing objects of art. In this summary definition he later also included the summary of Mellor (1922) on chemically induced physical damage caused by microorganisms on church windows. Krumbein (1998) went as far as to imply that one of the greatest German poets and scientists (namely Johann Wolfgang von Goethe) has selected the principle of chemical/physical change on rock surfaces as the theme of his novel "Die Wahlverwandschaften".

2. PATINA AS A GENERAL CHANGE OF THE SUBSTRATE OF OBJECTS OF ART

Biodeterioration of objects of art is only one aspect of the totality of changes the objects and the materials of which they are produced undergo. The other and possibly superimposed aspect is the development of a patina. Patina is the sum of all changes that the surface layers of objects of art are submitted to and inherit from environmental, biological and substrate related factors (Brachert, 1995; Toyka, 1996). This surface change (biodeteriorative or partially protective patina) is an exchange or biotransfer of material and energy between two heterogeneous open systems: The solid substrate (rock, concrete, mural painting, glass, wood, leather, parchment, paper etc.) and its environment (mainly atmosphere and, partially, hydrosphere). The exchange and biotransfer is certainly influenced largely by the bioreceptivity of the materials or substrates in question. The mutual interaction of all components and processes leads to a more or less complete turnover of the initial materials at the border between the two systems. This (bio-) transfer process may come to a standstill for certain periods of time when the conditions approach equilibrium (e. g. through patina, crusts etc.). It may, however, be revived if only one of the components or processes involved changes or is submitted to changes. In the total context of all interactions of organisms with the substrate, a patina can either be characterised by constant mass and composition of the substrate or by increase of mass. In both cases the changes will be less detrimental and possibly reversible. In the case of patina with loss of mass or physical coherence it turns into biocorrosive or biodeteriorative action. Biogenicity of patina on building stones and marbles (called also "scialbatura") was perhaps first introduced by Monte and Sabbioni (1987). Krumbein (1969) recorded biological influences on the formation of rock varnish a term which we now include into the embracing term, patina. Krumbein (1966) has first developed the above definition and modified it in the course of the expanding microbiological work in this field. Many data have been added by several authors. Detailed data on fungal detrimental chemical activity on sandstones (Braams, 1992) and chemical/physical activity on limestones and marbles (Gorbushina, 1997; Sterflinger, 1995) were presented.

Patina as an embracing term including biological impacts was put forward by Krumbein (1993). A patina may occur through changes of the morphology, the composition and the aesthetic values above, at or below the original material surface. The aesthetic values are usually expressed as changes of the spectral properties of the material (colour theory). Patina is one of the historically oldest and most embracing terms to characterise near surface changes of the material of objects of art (Boschini, 1660; Baldinucci, 1681). Therefore we assign priority to this term over other terms characterising the interactions of biota with culturally important substrates as e.g. crust, lacquer, rock varnish or any other kind of environmentally induced varnish formation, deposit, oxalate skin, efflorescence, deposit or others. Krumbein (1993) was not aware of our findings on biopigments in patination and of the consequences of morphological/mechanical changes of the surface leading to different spectral characteristics of the surfaces in question. Weathering or corrosion are not satisfactory terms to explain the total sum of all processes occurring between art substrates and the environment. Often the influences of physical, chemical and atmospheric agents is exaggerated. Ageing or patina formation describes the process of energy and material exchanges in a quasi-irreversible time-space system separated from reversible time-space systems by two circumstances:

1. patina formation changes irreproducible products of an individual artist;
2. patina formation in almost all cases is a biological process. Biological processes, in turn, are regarded as irreversible.

The formation of patina proceeds as an exchange between two heterogeneous, open, dynamic systems, namely:

1. the solid substrate (melts as e. g. glass; rocks as e. g. marble; artificial rock as e. g. mortar, concrete, fresco; half-organic material as e. g. ivory, paintings and fully organic materials as e. g. wax, tissue) and
2. its environment (internal and external atmosphere, water, neighbouring different materials etc.).

Both systems are characterised by

1. their innate physical characteristics (mass, particle size and form, volume, density, pressure, humidity, cohesion, diffusion constants, van der Waals and tunneling forces)
2. their organic/inorganic chemical, crystallographic, mineralogical, petrographic composition
3. their biological components and parameters (biophysical or mechanical, biochemical, physiological, evolutionary, ecological) and
4. their innate and added or subtracted energies (redox-status, sunlight, heat, biological phosphorylation or energy change).

The interaction of these factors leads to a continuous exchange of all components. An equilibrium with the lowest possible differences and gradients in energies and material characteristics is thermodynamically demanded by the total interactive system of organisms, substrate and environment. Speed and intensity of the exchange reactions are highest at the immediate borderline between the systems. The greater they are, the greater are the differences in the materials and their innate energy levels. A natural borderline to these exchange processes is defined by the penetration depth of all physical gradients, gases, solutions, organisms and their metabolic products. As a rule one can limit the environmental influences to a scale between several meters to a few mm and the interactions within the material to a maximum of a few cm and a minimum of some nm. The main characteristics of the slow ageing process leading to a patina as contrasted to rapid ageing leading to pitting, exfoliation, sanding, desquamation and total desintegration will be:

1. positive and, possibly, also negative fractal determined processes as well as
2. biochemical and epitaxial changes of neo-mineral formation or mineral destruction,
3. biologically increased factors of water pressure and changes in water density, and
4. the often biologically induced masking and redox modification of elements and ions leading to an acceleration of the biological-physical-chemical exchange processes.

The complex process of patina formation may come to a stand-still (still-life) when all these dynamic processes approach thermodynamic equilibrium. The ageing process, however, will invariably be renewed if only one of the multitude of factors of the three interfering systems is changed. The stand-still of the exchange processes is characterised by the development of a typical patina.

Patina can be regarded as a static or stabilising phase in the process of ageing or deterioration of an object of art.

In cases in which the patination process is mass-neutral, it is possibly also most stable. Stability, in turn, is most desirable for all objects of art.

One of the more recent biological contributions to the biostabilisation of surfaces of monuments may be the poikilotrophic principle (Gorbushina, 1997; Gorbushina et al., 1999) which states that the slowest growing microorganisms may out-compete the most detrimental organisms on monument surfaces under extremely changing or very extreme conditions for the growth of microorganisms. These conditions then should be created by conservators and restorers.

3. CHEMICAL CHANGES OF SUBSTRATES

Chemical changes of the substrates which have been used and are used for the production of objects of the cultural heritage have been analysed in many scientific investigations and have been the topic of many scientific publications. In the case of monuments acid rain and urban pollution have been mentioned many times and analysed by all possible means (e.g. Baer et al., 1991; Zezza, 1991; Thiel, 1993; Baer and Snethlage, 1997). In this context, it suffices to state that any chemical change of an object under surface conditions may always be influenced by the chemical changes accelerated or retarded by biological interaction (Mellor, 1922). One example may be the interaction of chemolithoautotrophic microorganisms with the continuously changing compounds of the nitrogen and sulphur cycles under formation of nitric and sulphuric acid. This knowledge stems from workers like Kauffmann (1953) for the nitrogen cycle or Isacenko (1936) for the sulphur cycle. A summary was given by Bock and Krumbein (1989). Both authors hinted that biogenic nitric and sulphuric acids may largely enhance the decay of many objects of art and buildings. Krumbein (1998) summarised some aspects of interactive chemolithotrophic and chemoorganotrophic acid production. Alexandrov (1949) and Krassilnikov (1951) have already largely contributed to the factors of chemoorganotrophic activity in the decay of the rock substrate. This chemical degradation often is a chemical process leading to neomineral formation which, in turn, initiates physical or mechanical damage. Recently, Banfield and Nealson (1997) collected a large body of information concerning these chemical-physical-mineralogical relationships. Moreover, organic compounds produced through the metabolic activity of microorganisms (acids and bases) may often alter the substrate faster than inorganic acids produced simultaneously.

Thus major factors of a chemical conditioning of a given substrate are represented by microbial metabolic activities and products excreted into the environment. The organisms settle on and within the substrate. They may or may not use parts of the substrate as nutrients or energy sources. They may receive all nutrients and energy sources from the surrounding atmosphere and by wet and dry deposition. The metabolic products may be accumulated first in the cells and only later be excreted into the environment upon the death and lysis of the cells. On the other hand, the uptake of nutrients as well as the use of external electron donors and acceptors will yield oxidation-reduction products which may drastically change the rock substrate during the lifetime of a sub-aerial microbial biofilm (Gorbushina and Krumbein, in press, b). One logical consequence is the permanent excretion of inorganic and organic acids and bases and the direct dissolution of parts of the substrate.

Another feature is the production of chelating and buffering substances which change the whole sequence of substrate-environment interactions. Some chemical changes of the substrate may not be directly visible because the end products are washed out by rain. Others may transform parts of the substrate or add layers of freshly precipitated minerals. A very special case in most rock surfaces is the biologically-induced and catalysed transformation of calcium carbonate and silicate into calcium oxalates. The oxalates represent highly insoluble and also biologically refractory minerals which accumulate in the form of patina or crust on the surface and within the surficial layers of any rock monument. The overall (bulk) chemical composition of the substrate may not be changed this way. However, these biologically-induced chemical changes will lead to serious physical changes of the substrate. Very useful and detailed reviews of the microbiological impact on chemical changes of substrates of objects of art have been given by Urzi and Krumbein (1994) and Koestler et al. (1997).

4. PHYSICAL CHANGES OF SUBSTRATES

The question of physical changes of the substrate by abiogenic forces has attracted attention already for quite some time (Ginell, 1994). However, microbially-induced or produced physical and mechanical changes of the substrate of objects of art have been treated so far only accidentally and ephemerally. A great gap exists in our knowledge concerning the biological physical/mechanical impact on the substrate of objects of art. Microorganisms have, however, an important physical impact on substrates used for objects of art which can be derived from basic physical laws and mechanisms. The most vulnerable substrates in this case seem to be rocks and glasses. Organic tissues are very bioreceptive, while wood, parchment and paper seem to be more resistant than stone.

The strongest physical actions on substrates are probably caused by:

1. vibrations, shocks and so-called disasters
2. spectral changes
3. thermal energy transfer
4. mechanical action of salts
5. mechanical action of water, liquid and vapours
6. mechanical action of living cells, tissue, membranes and molecules
7. interference of small organics with internal (mineral) surfaces
8. electrostatic effects
9. gas diffusion and its consequences.

All physical or rather mechanical actions of microorganisms are, evidently, at the level of the microscopic scale in biological terms or the fractal scale in physical terms and need to be followed with techniques, which in cases are not yet available. Wiggins (1990) has expressed the strong feeling that the mechanical power of water in biological systems and extracellular polymeric substances is tremendous. She added that, presently, we have no equipment to measure these biogenic physical forces, but we observe them.

Many other physical factors in the microbe-substrate interactions have been overlooked so far or have been studied very rudimentarely. In a research project on the molecular aspects of contamination control of mural paintings (MICROCORE) of the EEC, we studied the physical parameters of growth of microorganisms and the molecular biology of the rock inhabiting flora using the biodiversity approach of Amann et al. (1995), Roelleke et al. (1996) and Sterflinger and Gorbushina (1997). Since the eukaryote signals are also a subject in this study more attention was given to stone-lice and other insects inhabiting the cultural objects. It was found that not only the grazing action of the insects but also their transfer activity of spores and propagules cannot be underestimated as a physical factor.

Biopitting and exfoliation through physical and mechanical actions of microbiota have been demonstrated by Danin (1992), Gehrmann (1995), Sterflinger (1995), Sterflinger and Krumbein (1997) and Gorbushina (1997). Sun and Friedmann (1999) have shown independently that microbiota on and in the rock substrate may physically alter the material on geological time scales. Similar ideas have already been put forward by Bachmann (1917) and Krumbein (1969).

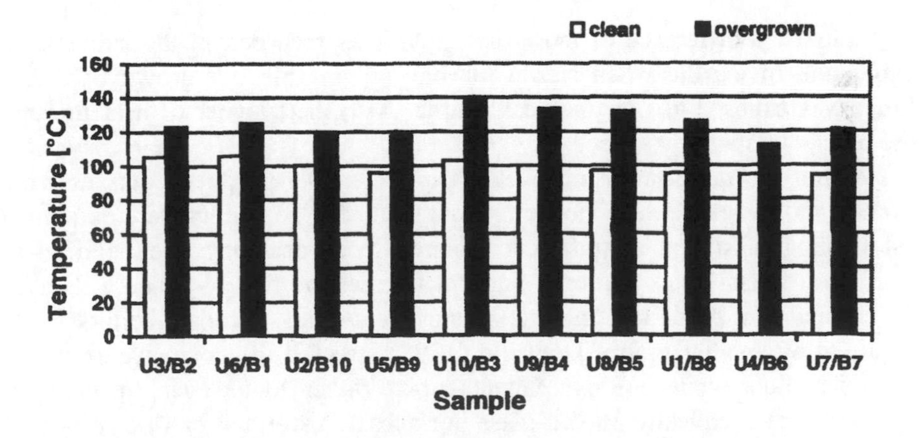

Figure 1. Temperatures of several different marble blocks when clean and sterile (white) and when overgrown by poikilotrophic black fungi (black). Overgrown blocks heat up much faster than sterile ones. Also the heat transfer through the blocks differs considerably.

Recently, in a programme of the Physics Department and the ICBM of the Carl von Ossietzky University, two of us (Gorbushina and Dornieden) have initiated a series of experiments on mechanical impacts of microorganisms on rock surfaces (mainly marble) and glass (Dornieden et al., 1997, Gorbushina and Palinska, in press). Poikilotrophic fungi were inoculated under controlled environmental conditions in the laboratory on marble and glass blocks and between marble or glass slices in order to verify their mechanical impact on the substrate. Fig.s. 1 and 2 demonstrate initial results in terms of thermal behaviour and dilatation of marble when infected by poikilotrophic fungi.

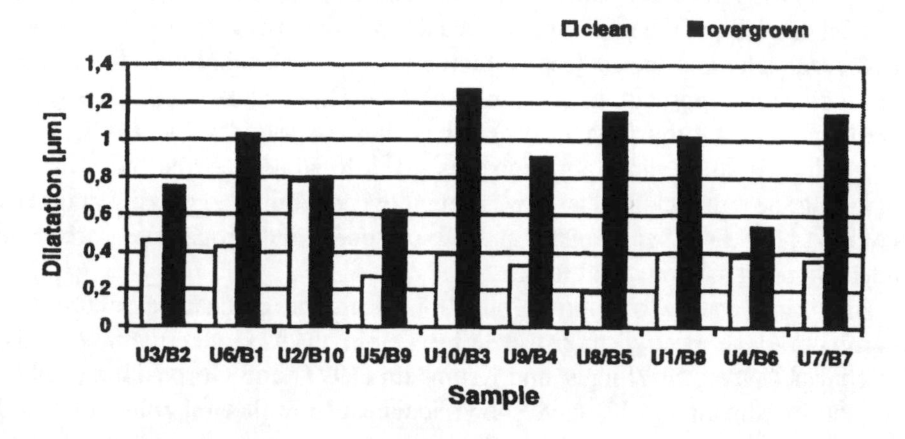

Figure 2. Dilatation of different marble blocks when clean and sterile (white) and when overgrown by poikilotrophic black fungi (black). The (irreversible) dilatation values are much higher in the case of the infected marble samples.

Actually a difference of more than 20% was recorded in the irreversible expansion of marble when sterile marble and marble overgrown by black fungi was exposed to a strong heat source. Also the transfer of heat through the blocks was largely altered.

Within the same programme also experimentally-produced glasses from France and Germany were inoculated and studied. The results were extremely astonishing inasmuch as different isolates from decaying rock and glass caused different crack patterns and thicknesses of the so-called gel layer developing on biodeteriorating glass surfaces. Very similar results were obtained by another group. Drewello (1998) stressed the corrosive action of the microbiota while our experiments point rather to a severe mechanical attack of the chemically altered glass surfaces. This attack can be compared to the mechanical influences of lichens on the rock substrate postulated by Bachmann (1917) and demonstrated by Gehrmann (1995) and Nimis and Tretiach (1995).

Figure 3. Fungal growth inside and on marble indicating mechanical stress. Microorganisms may exert strong mechanical forces on the surrounding mineral particles and the adhesion of minerals to each other. 27177 shows a fungal colony within a cavity; 27182 demonstrates the mechanical splitting of marble grains; 27197 shows a colony within a natural (?) crack; 27178 fungal cells and hyphae penetrating (actively?) the stone.

To our astonishment also the species of fungi were important for the physical and morphological pattern of cracks forming on glass surfaces (Gorbushina and Palinska, in press). Drewello (1998) worked mainly with strains from collections or ephemeral typical soil fungi. In our experiments, poikilotrophic fungi of the typical habitat were selected and tested.

Doubtlessly, the depth of the gel layer as well as the pattern and density of cracks formed during the growth process were closely related to the fungal strains isolated from rock or glass and used for the subsequent laboratory experiments. Interestingly, the work on crust formation and corrosion of ancient glass by Drewello (1998) supported the finding of Gorbushina et al. (1995) that biogenic formation of gypsum occurs on glass and marble. Thus even the well studied black gypsum crusts on monument and ancient glass surfaces have been shown independently by two groups as being at least in part produced by a biological reaction chain and not only through air polluted with sulphur dioxide as believed for many years. Research and new findings on biogenic chemical, mineralogical and physical interactions with detrimental or protective influences on the substrates of objects of art thus still have an important place in the basic research on conservation and restoration of objects of art.

5. CONCLUSIONS

Physical or mechanical abrasion, decohesion and spectral changes may be the most important factors in the processes leading to patina, biopitting and exfoliation or desquamation. The recent work of Dornieden et al. (1997) and Drewello (1998) as well as the very early works of Bachmann (1917) and Mellor (1922) and the embracing biogeochemical and biogeophysical views of Vernadsky (1997) contributed largely to the following concluding remarks.

The study of the chemical and physical relationships of microbes, substrate and environment in the case of the decay and conservation of objects of art submitted to patina formation or rapid decay inevitably leads to the following summarising remarks:

1. The immediate cause of the destruction of objects of art is a mechanical action exerted by the biofilms establishing themselves on chemically pre-conditioned surfaces and substrates.
2. The chemical corrosion or pre-conditioning of the substrate is always accelerated by sub-aerial microbial biofilms.
3. Inhibitory substances such as heavy metals or organic antagonists may cause exceptional resistances of these substrates.
4. Annual careful and cautious cleaning of the substrate surfaces may help to avoid the physical/mechanical attack and destruction of the substrates because the establishment of a mature sub-aerial aggressive biofilm needs a considerable period of time which is definitely longer than in the case of sub-aquatic detrimental biofilms (Mellor, 1922).

ACKNOWLEDGEMENTS

This work was supported by MICROCORE (project Nr. ENV4-CT97-0705, Novel molecular tools for the analysis of unknown microbial communities of mural paintings and their implementation into the conservation/restoration practice) of the European Commission and by a grant on glass and marble decay by physical/mechanical and chemical influences of microorganisms from the Deutsche Forschungsgemeinschaft, project Kr 333/25-1/2.

REFERENCES

Alexandrov, V. G. 1949. Bacteria destroying silicates. Mikrobiologiya **19:** 97-107.

Amann, R. I., Ludwig, W. and Schleifer, K.-H. 1995. Phylogenetic identification and in situ detection of individual microbial cells without cultivation. Microbiol. Rev. **59:** 143-169.

Bachmann, E. 1917. Die Beziehungen der Kieselflechten zu ihrem Substrat. Ber. Dtsch. Bot. Gesellsch. **35:** 464-476.

Baer, N. S., Sabbioni, C. and Sors, A. I. (eds.) 1991. Science, Technology and European Heritage. Butterworth-Heinemann, Oxford.

Baer, N. S. and Snethlage, R. (eds.) 1997. Saving our architectural heritage. The conservation of historic stone structures. Wiley, Chichester.

Baldinucci, F. 1681. Vacabulario de la lingua toscana, Firenze.

Banfield, J. F. and Nealson, K. H. (eds.) 1997. Geomicrobiology: Interactions between microbes and minerals. Reviews in Mineralogy Volume 35. Mineralogical Soc. of America, Washington D. C.

Bock, E. and Krumbein, W. E. 1989. Aktivitäten von Mikroorganismen und mögliche Folgen für Gestein von Baudenkmälern. Bautenschutz Bausanierung, Sonderheft p. 34-37.

Boschini, M. 1660. La carta del Navigar pittoresco. Venezia.

Braams, J. 1992. Ecological studies on the fungal microflora inhabiting historical sandstone monuments. PhD thesis, Oldenburg.

Brachert, Th. 1995. Patina. Callwey, München.

Danin, A. 1992. Biogenic weathering of marble monuments in Didim, Turkey, and in Trajan's column, Rome. *In* Proc. 5th Intern. Conf. On Environm. Quality and Ecosystem Stability, Jerusalem p. 675-681.

Diakumaku, E. 1995. Investigations on the role of black fungi and their pigments on the deterioration of monuments, Ph.D Thesis, Oldenburg.

Diakumaku, E., Gorbushina, A. A., Krumbein, W. E., Panina, L. and Soukharjevski, S. 1995. Black fungi in marble and limestones - an aesthetical, chemical and physical problem for the conservation of monuments. Sci. Total Environm. 167: 295-304.

Dornieden, Th., Gorbushina, A. A. and Krumbein, W. E. 1997. Changes of the physical properties of marble as a result of fungal growth. Int. Journal for Restoration of Buildings, **3:** 441-456.

Drewello, R. 1998. Mikrobiell induzierte Korrosion von Silikatglas - unter besonderer Berücksichtigung von Alkali-Erdalkali-Silikatgläsern. PhD thesis, Erlangen.

Gehrmann, C. I. 1995. On the biopitting corrosion by epilithic and endolithic lichens on carbonate rocks - biophysical and biochemical weathering aspects. PhD thesis, University of Oldenburg.

Ginell, W. S. 1994. The nature of changes caused by physical factors. *In* Durability and Change. The science,responsibility and cost of sustaining the cultural heritage, Krumbein, W. E., Brimblecombe, P., Cosgrove, D. E. and Staniforth S. (eds.), Wiley, Chichester p.81-94.

Gorbushina, A. A. 1997. Biological properties of marble deteriorating fungi, Ph. D Thesis, St. Petersburg.

Gorbushina, A. A. and Krumbein W. E. 1999a. Poikilotroph response of microorganisms to shifting alkalinity, salinity, temperature and water potential. *In* Microbiology and biogeochemistry of hypersaline environments, Oren A.(ed.), CSC, Boca Raton, Florida p. 75-86.

Gorbushina, A. A. and Krumbein, W. E. 1999b. The poikilotrophic microorganism and it's environment. Microbial strategies of establishment, growth and survival. *In* Enigmatic microorganisms and life in extreme environments, J. Seckbach (ed.), Kluver, Dordrecht p 177-185.

Gorbushina, A. A. and Krumbein, W. E. (in press, a). Rock dwelling fungal communities: diversity of life styles and colony structure. *In* Microbial Diversity, Seckbach, J. (ed.), Kluwer, Dordrecht.

Gorbushina, A. A., Krumbein, W. E., (in press, b). Microbial effects on sub-aerial rock surfaces. *In* Microbialites, Riding, R. and Awramik, S. (eds.), Springer, Berlin.

Gorbushina, A. A., Krumbein, W. E., Hamman, C. H., Panina, L., Soukharjevski, S. and Wollenzien, U. 1993. Role of black fungi in colour change and biodeterioration of antique marbles. Geomicrobiol. Journ. **11:** 205-222.

Gorbushina, A. A., Krumbein, W. E. and Palinska, K. A. 1999. Poikilotroph growth patterns in rock inhabiting cyanobacteria. *In* The Phototrophic Prokaryotes, Peschek et al. (eds.), Kluwer Academic/Plenum Publishers, N. Y. p 657-664.

Gorbushina, A. A., Krumbein, W. E. and Vlasov, D. Y. 1995. Biocarst cycles on monument surfaces. *In* Conservation and restoration of cultural heritage, Pancella, R. (ed.), Proc. of the 1995 LCP Congress, Ecole Polytechnique Federale, Lausanne p. 319-332.

Gorbushina, A. A. and Palinska, K. A. Aerobiology, in press.

Gorbushina, A. A., Panina, L. K., Vlasov, D. Y., Krumbein, W. E. 1996. Fungi deteriorating Chersonesus marble. Mikologia i Fitopatologia **30:** 23-28.

Gromov, B. V. 1963. Microflora of deteriorating brick, mortar and marble. Vestnik LGU **3:** 69-77.

Guilitte, O. 1994. Bioreceptivity: a new concept for building ecology studies. The Science of the total Environment. **167:** 215-250.

Isacenko, V. 1936. Sur la corrosion du beton. Doklady Acad. Sci. USSR **II:** 287-289.

Kauffmann, J. 1953. Role des bacteries dans l'alteration des pierres des monuments. Corrosion et Anticorrosion **1:** 33-41.

Koestler, R. J., Warscheid, Th. and Nieto, F. 1997. Biodeterioration: Risk factors and their management. *In* Saving our architectural heritage. The conservation of historic stone structures, Baer, N. S. and Snethlage, R. (eds.), Wiley, Chichester p.25-36

Krassilnikov, B., 1951. Role des microorganismes dans l'erosion des differentes types de roches. Mikrobiologiya, **20:** 90-98.

Krumbein, W. E. 1966. Zur Frage der Gesteinsverwitterung (Über geochemische und mikrobiologische Bereiche der exogenen Dynamik). Inauguraldissertation, Universitaet Wuerzburg.

Krumbein, W. E. 1969. Ueber den Einfluß der Mikroflora auf die exogene Dynamik (Verwitterung und Krustenbildung). Geol. Rdsch. **58**: 333-363.

Krumbein, W. E. 1993. Zum Begriff Patina, seiner Beziehung zu Krusten und Verfärbungen und deren Auswirkungen auf den Zustand von Monumenten. *In* Steinzerfall-Steinkonservierung, Snethlage, R. (ed.), Ernst und Sohn, Berlin p. 215-229.

Krumbein, W. E., 1998. Mikrobenbefall und Steinzerstörung: autotroph oder heterotroph? chemisch oder physikalisch? Strategien der Verhinderung und Behebung - Eine Bilanz. *In* Denkmalpflege und Naturwissenschaft. Natursteinkonservierung II, Snethlage R. (ed), Fraunhofer IRB Verlag, Stuttgart p. 173-205.

Mellor, E. 1922. Les lichens vitricoles et la détérioration des vitraux d'église. Thèse de Doctorat, Sorbonne, Paris.

Monte, M. del and Sabbioni, C. 1987. A study of patina called "scialbatura" on imperial Roman marbles. Studies in Conservation **32**: 114-121.

Nimis, P.-L. and Tretiach, M. 1995. Studies on the biodeterioration potential of lichens, with particular reference to endolithic lichens. *In* Interactive weathering and bioreceptivity study on building stones, monitored by computerized X-ray tomography (CT) as a potential non-destructive research tool. Research report 2 of the Programme on Protection and Conservation of the European cultural heritage, deCleene, M. (ed.), Science information office, University of Gent, Gent p. 63-122.

Roelleke, S., Muyzer, G., Wawer, KL., Wanner, G. and Lubitz, W. 1996. Identification of bacteria in a biodegraded wall painting by denaturing gradient gel electrophoresis of PCR-amplified gene fragments coding for 16S rRNA. Appl. Environm. Microbiol. **62**: 2059-2065.

Staley, J. T., Palmer, F. and Adams, J. B. 1982. Microcolonial fungi: Common inhabitants on desert rocks. Science **215**: 1093-1095.

Sterflinger, K. 1995. Geomicrobiological investigations on the alteration of marble monuments by Dematiaceous fungi (Sanctuary of Delos, Cyclades, Greece). PhD thesis, Oldenburg.

Sterflinger, K. and Gorbushina, A. A. 1997. Morphological and molecular characterization of a rock inhabiting and rock decaying dematiaceous fungus isolated from antique monuments of Delos (Cyclades, Greece) and Chersonesus (Crimea, Ukraine). Syst. Appl. Microbiol. **20**: 329-335.

Sterflinger, K. and Krumbein, W. E. 1997. Dematiaceous fungi as a major agent for biopitting on mediterranean marbles and limestones. Geomicrobiol. J. **14**: 2199-230.

Sun, H. J. and Friedmann, E. I. 1999. Growth on geological time scales in the antarctic cryptoendolithic microbial community. Geomicrobiol. J. **16**: 193-202.

Thiel, M.-J. 1993. Conservation of stone and other materials. E&FN Spon, London.

Toyka, R. 1996. Patina. Junius Verlag, Hamburg.

Urzi, C. and Krumbein, W. E. 1994. Microbiological impacts on cultural heritage. *In* Durability and Change. The science, responsibility and cost of sustaining the cultural heritage, Krumbein, W. E., Brimblecombe, P., Cosgrove, D. E. and Staniforth S. (eds.), Wiley, Chichester p.107-135.

Vernadsky, V. I. 1997. Der Mensch in der Biosphäre. Europäischer Verlag d. Wiss., Frankfurt.

Wiggins, P. M. 1990. Role of water in some biological processes. Microbiol. Reviews **54**: 432-449.

Zezza F. 1991. Weathering and air pollution. Mario Adda, Bari.

A LABORATORY INVESTIGATION OF THE MICROBIAL DEGRADATION OF CULTURAL HERITAGE

AnnaMaria Seves[1], Maria Romanò[2], Tullia Maifreni[2], Alberto Seves[3], Giovanna Scicolone[4], Silvio Sora[1] and Orio Ciferri[1]

[1]Department of Genetics and Microbiology "A. Buzzati Traverso", University of Pavia, via Abbiategrasso 207, I-27100 Pavia, Italy; [2]Stazione Sperimentale per la Seta, via G. Colombo, I-20133 Milano, Italy; [3]Stazione Sperimentale per la Cellulosa, Carta e Fibre Tessili Vegetali ed Artificiali, Piazza L. da Vinci 26, I-20133 Milano, Italy; [4]Scuola Regionale per la Conservazione dei Beni Culturali, I-25082 Botticino Sera (BS), Italy.

Key words: paintings, silk, microbial degradation

Abstract: The effect of aging on the microbial colonization of two model systems, "mock paintings" and silk, has been investigated in samples artificially aged. In the case of mock paintings, as compared to untreated controls, aging by heat treatment increases colonization by the two tested fungal species (almost two orders of magnitude in the most aged samples). Relining of aged paintings seems to result in an increase in the colonization by bacteria and fungi if wheat starch paste is used in the relining procedure. In the case of silk fibroin, aging by treatment with wet heat or exposure to a Xenon lamp does increase to a small extent the susceptibility to bacterial invasion. However, fungi seem to grow on artificially-aged fibroin but not on the untreated protein. Thus it appears likely that the chemical modifications induced by the two treatments render fibroin a source of carbon and nitrogen utilisable also by fungi. The latter data may be of some relevance to the conservation of silk artifacts of historical or artistic interest.

1. INTRODUCTION

It seems to me that we should now realise that studies of the role of microorganism in the defacement and degradation of cultural heritage must go beyond the descriptive stage, that is cataloguing which organisms are found on which substrate. Certainly, this type of information is important since to establish which organisms colonize a given art work is necessary for any corrective or disinfestation treatment. However, I think that we can now begin to study and understand the mechanisms underlying the microbiological attack by setting up standardised laboratory models utilizing the most common types of support as well as the most commonly employed ingredients. These models will allow to establish, under controlled conditions, which species colonize a given substrate, how the microbial flora will change on changing the components of an art work (supports, pigments, binders, glues, etc.), how the substrate is modified by the microbial colonization and how these modifications lead to the establishment of different microbial communities. Similarly, one should try to evaluate in the laboratory how the microbial population varies when the environmental conditions change. Finally, one must assess how aging, which may be simulated in the laboratory, may bring about variations in the chemical structure of many components of works of art and how these differences often bring about colonization by different microbial taxa.

A few laboratory models have been already devised and utilized to test the contribution of microorganisms to the degradation of cultural heritage. For instance, the role of thiobacilli in the sulphuric acid degradation of concrete and of nitrifying bacteria in the case of sandstone has been established in controlled laboratory experiments (Bock and Stone, 1993). Similarly, sterilized samples of limestone inoculated with pure cultures of a cyanobacterium or a green alga showed, after 2 months of incubation, weathering and loosening of stone grains similar to those observed in field samples (Ortega-Calvo et al., 1991). In the case of paintings, linen canvases or wooden panels were grounded and different binding materials and pigments applied to them (Wazny and Rudniewski, 1972). These samples were placed on petri plates containing agarized nutrient medium inoculated with spore suspensions of six different fungi. After two weeks of incubation, the extent of fungal growth and sporulation was determined to ascertain which of the supports, binding materials and pigments were more resistant to fungal attack. In another set of experiments, wood panels coated with a white acrylic paint were exposed to soil. After a suitable time of incubation, representatives of seven genera of bacteria and fifteen of fungi were isolated from such samples (O'Neil, 1986). Whereas the population of most bacterial species remained constant or increased only slightly during the duration of the experiments, that of *Pseudomonas* spp. increased linearly with the time of incubation (during 12 weeks of incubation the number of *Pseudomonas* colonies per cm^2 increased by more than one order of magnitude). In the case of fungi, only *Aureobasidium (Pullularia) pullulans*, considered by some as the main biological agent of paint deterioration (Reynolds, 1950), increased steadily with the time of incubation. Indeed, after 12 weeks of incubation, *A. pullulans* was essentially the only fungal species present on the panels confirming an earlier report on the succession of fungi on this type of substrate (Winters et al., 1975).

In our laboratory, we have developed and utilized two models for evaluating the colonization of works of art by the microbial flora of soil. One model relies on "mock paintings", essentially linen canvas sized with animal glue in water on which a paint film of lead white is laid (Seves et al., 1995; Seves et al., 1996). The other one, is a proteinaceous substrate, raw or degummed silk (Seves et al., 1998). For both substrates, exposure to soil or soil extracts has allowed the isolation and identification of the quantitatively most important microbial species growing on such substrates. The main bacterial and fungal species isolated in these experiments (Table) are utilised to investigate the extent and the effects of the microbial colonization of the two model systems.

Microorganisms utilized

	Bacteria	**Fungi**
Paintings	*Bacillus amyloliquefaciens*	*Aspergillus niger*
	B. pumilus	*Penicillium chrysogenum*
	Pseudomonas maltophilia	
	Variovorax paradoxus	
Silk	*Arthrobacter aurescens*	*Aspergillus niger*
	Bacillus megaterium	*Cladosporium cladosp.*
	Chryseomonas luteola	*Penicillium chrysogenum*
	Variovorax paradoxus	

2. EXPERIMENTAL RESULTS

Aging of "mock paintings" (Seves et al., 1999) does not influence significantly their susceptibility to colonization by pure cultures of the soil bacteria (Fig.. 1) that were found previously to be the most common colonizers of these substrates. In these experiments the species that grew more abundantly were *B. amyloliquefaciens* and *Ps. maltophilia* whereas *V. paradoxus* grew to a limited extent and *B. pumilus* did not grow at all. In the case of the two fungi, laboratory-aged "mock paintings" appeared to be colonized by both species more rapidly and extensively that untreated samples (Fig.. 1). The differences seem to be significant since, as compared to untreated controls, fungal CFU's were approximately one order of magnitude higher in samples aged the equivalent of 135 and 179 years

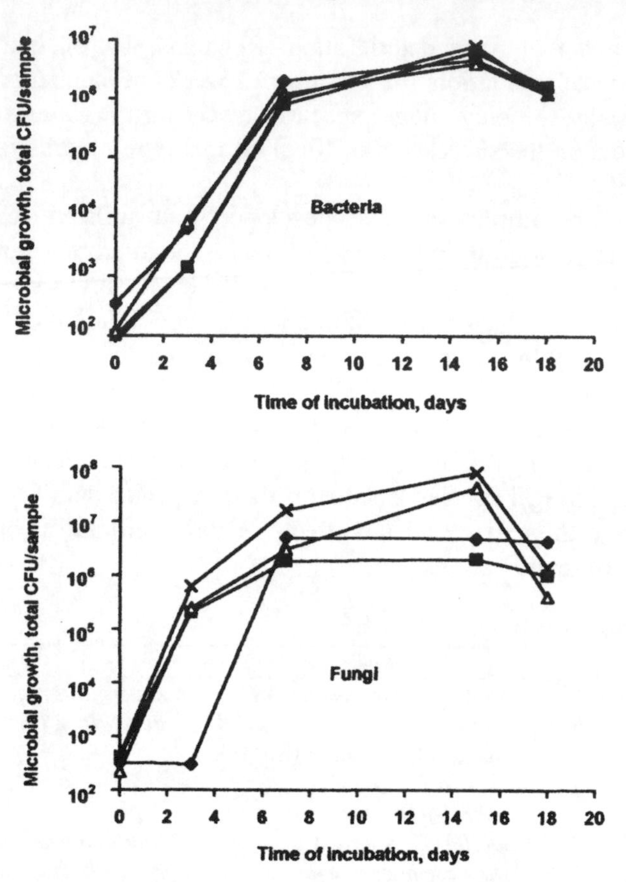

Figure 1. Growth of bacteria (top) and fungi (bottom) on laboratory-aged "mock paintings". "Mock paintings" were heat-treated at 105°C in the dark for periods of time varying from 144 to 700 hours corresponding to natural aging of 37 to 179 years. Control (no heat treatment) ◆-◆, aging equivalent 37 years ■-■, 135 years **x-x**, 179 years Δ-Δ.

An important operation that is performed on old paintings is their relining through the gluing of a new, in general canvas, support. Of the three glues most commonly utilized in this process, Beva (ethyl-vinyl acetate), Plextol (polymetacrylate) and wheat starch paste, only the latter stimulated significantly the rate and the extent of both bacterial and fungal colonization (Fig.. 2). If laboratory-aged "mock paintings" were re-lined and then aged for a further equivalent of 50 years, microbial colonization appeared to be much more abundant (one order of magnitude higher than that observed if aging was not imposed after relining) (Fig.. 3). In these experiments wheat starch paste stimulated microbial growth slightly more than other glues and the stimulation was most evident in the case of fungi. Finally, the addition of a fungistatic agent, Na o-phenylphenate (0.3%, w/w), commonly added to glues used for relining, had a rather limited effect since it decreased only slightly fungal colonization (less than one order of magnitude compared to controls) and, more important, delayed of just a few days the onset of fungal colonization (Fig.. 4). Thus addition of the fungistatic compound may be of some advantage only in the case of paintings that, after relining, are kept under controlled environmental conditions that prevent microbial growth (e.g. in museums). In conclusion, these data indicate that, as expected, aged paintings are somewhat more susceptible to microbial colonization and that relining increases this susceptibility especially if a glue containing wheat starch, an excellent source of nutrients, is utilized. If a relined painting is further aged, then its susceptibility to colonization by microorganisms is further enhanced.

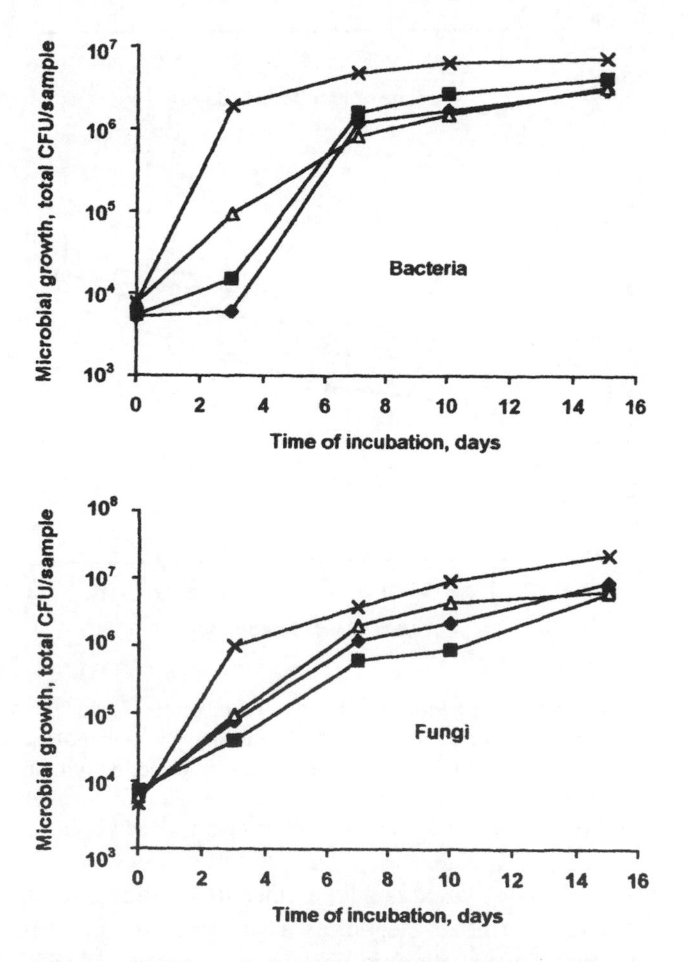

Figure 2. Growth of bacteria (top) and fungi (bottom) on laboratory-aged "mock paintings" relined with different glues. "Mock paintings" were heat-treated for a period of time corresponding to a natural aging of 100 years ◆-◆ and then relined with a linen canvas utilising as a glue Beva ■-■, Plextol Δ-Δ or wheat starch paste x-x.

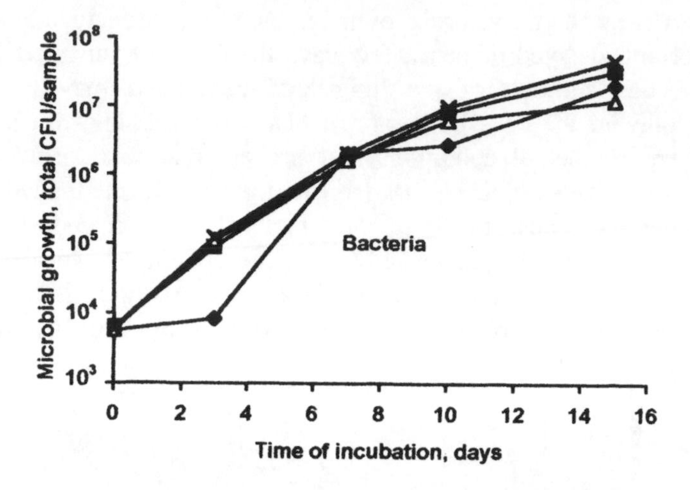

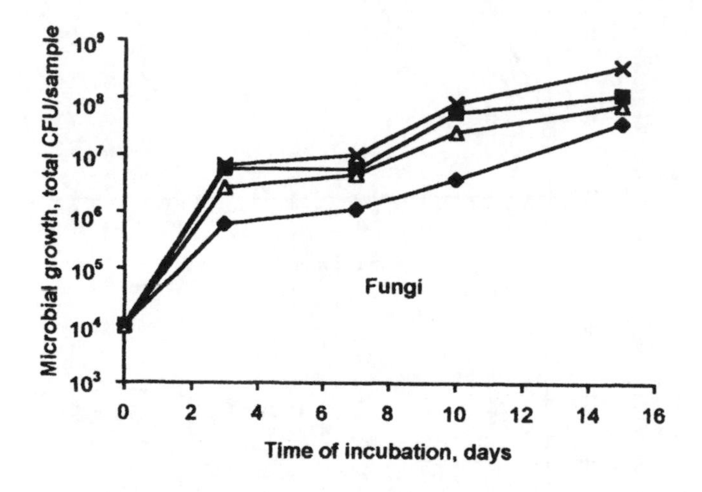

Figure 3. Growth of bacteria (top) and fungi (bottom) on laboratory-aged "mock paintings" relined and aged for a further period of time. "Mock paintings" were heat-treated and relined as in Fig.. 2 but, after relining, heat-treated for a further period of time corresponding to a natural aging of another 50 years. "Mock paintings" not relined ◆-◆, relined with Beva ■-■, relined with Plextol Δ-Δ, relined with wheat starch paste **x-x**.

The other model investigated is silk, a fiber utilized as such for art works (e.g. tapestries, rugs, garments etc.) or as a substrate for paintings or prints. We previously reported that the two protein components of raw silk, sericin and fibroin, may be utilized as a source of carbon and nitrogen for growth by many bacteria but not by fungi present in soil (Seves et al., 1998). (However, very recently we have isolated a species of *Verticillium* which seems capable to grow in a mineral medium in which fibroin is the only source of carbon and nitrogen). Soil burial of fibroin severely reduces the mechanical properties of the protein resulting in considerable reduction of viscosity, elongation, and breaking load values. On fibroin, bacteria appear to form a well-developed biofilm in which bacterial cells (and possibly other soil microorganisms) are tightly bound, probably through the production of extracellular polymeric substances (Fig.. 5).

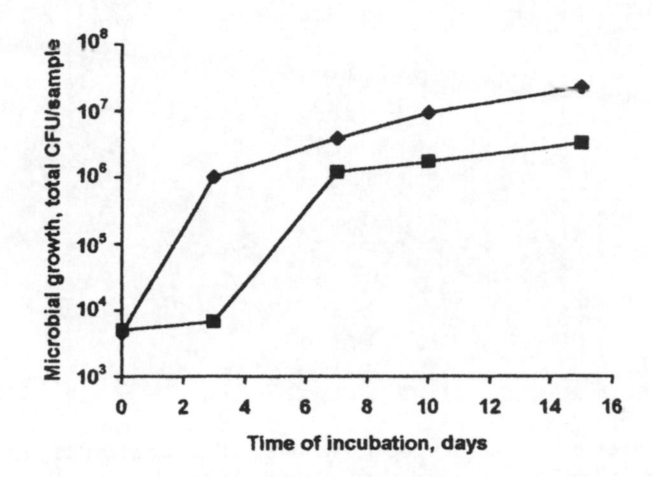

Figure 4. Effect of the addition of a fungistatic agent on growth of fungi on "mock paintings". "Mock paintings" were aged in the laboratory for the equivalent of 100 years and then relined utilizing wheat starch paste only ◆-◆ or additioned of 0.3% (w/w) Na o-phenylphenate ■-■.

We now wish to report the susceptibility to microbial attack of artificially-aged fibroin. Aging has been achieved by exposing fibroin samples to wet heat or a Xenon light, treatments that reduce tensile strenght and the other mechanical properties of the polymer (Kurupillai et al., 1989). Heat treatment increases only to a certain extent fibroin's susceptibility to colonization by selected bacterial species (*Arthrobacter aurescens, Bacillus megaterium, Chryseomonas luteola, Variovorax paradoxus, A. aurescens* and *V. paradoxus* being the most abundant colonizers) (Fig.. 6). However, aging renders the protein susceptible also to colonisation by fungi. Indeed, the three fungal species tested do not grow at all on controls but they appear to colonize fibroin heat-treated for periods of time resulting in aging equivalent to 6 and 10 years of storage. If aging is accomplished by exposure to a Xenon light lamp, fibroin's colonization by bacteria is essentially not affected by the treatment but fungi appear to grow quite well on the protein exposed to light (Fig.. 7). It may be surmised that the chemical and physical degradation of fibroin brought about by the two treatments renders the protein utilizable as a source of carbon and nitrogen for growth even by fungi.

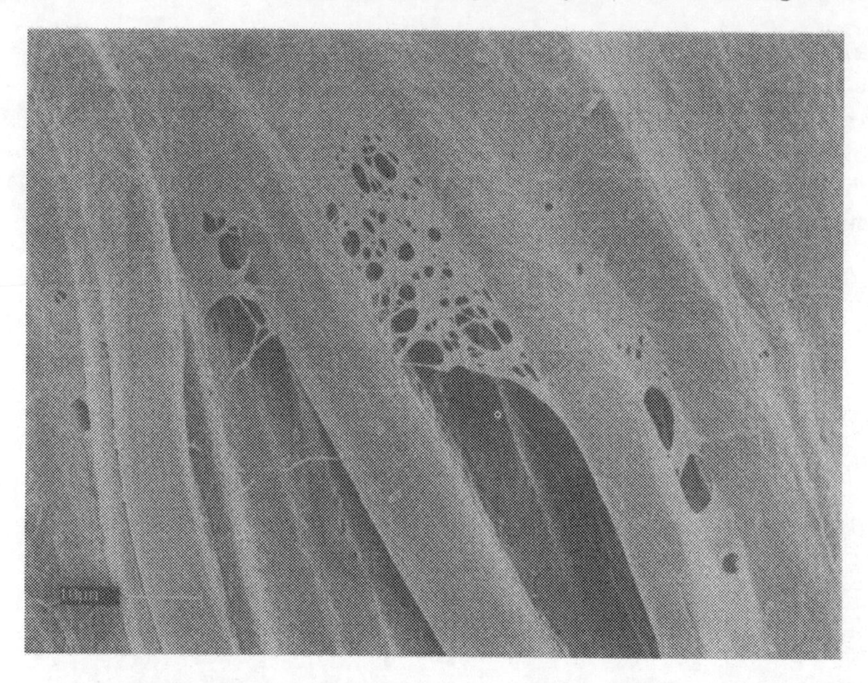

Figure 5. SEM of the microfilm developing on fibroin's threads incubated with four cultures of soil bacteria. A few isolated bacterial cells are visible as well as many bacterial clusters covered by a microbial biofilm.

These and the previously reported results indicate that raw silk (a mixture of sericin and fibroin) but also degummed silk (fibroin only) may be substrates on which certain soil microorganisms may adhere to and grow onto reducing the mechanical properties of the proteins. Some of these microorganisms appear to produce a biofilm on the fiber's threads. This biofilm may entrap also other microbial species that may utilize for growth the products of fibroin's degradation as well as the metabolites of the species degrading the protein. Further, fungi appear to be capable of growing on fibroin artificially aged (Fig.. 8) and, therefore, it is quite likely that they may participate to the biodegradative community that bacteria establish on silk's threads. These and previous results indicate that, under suitable environmental conditions, silk may be degraded and utilized by microorganisms as a source of carbon and nitrogen. Finally, although we have no data in the case of naturally-aged silk, the results obtained with fibroin artificially-aged indicate that old silk artifacts of historical or artistic interest may be prone to colonization and hence to degradation not only by bacteria but also by fungi.

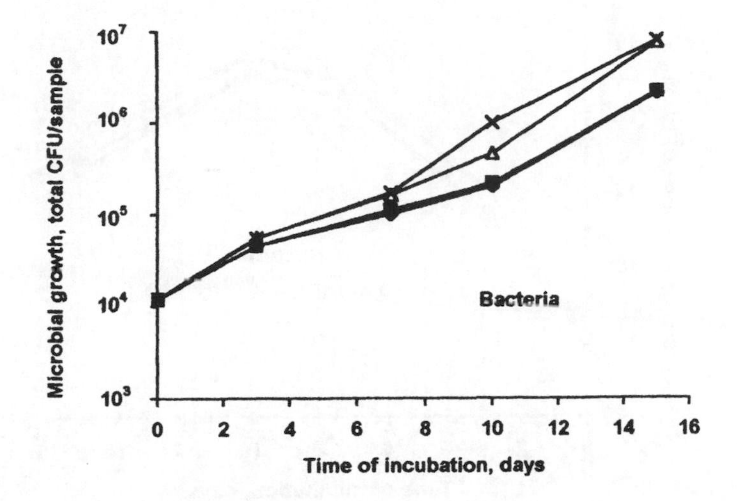

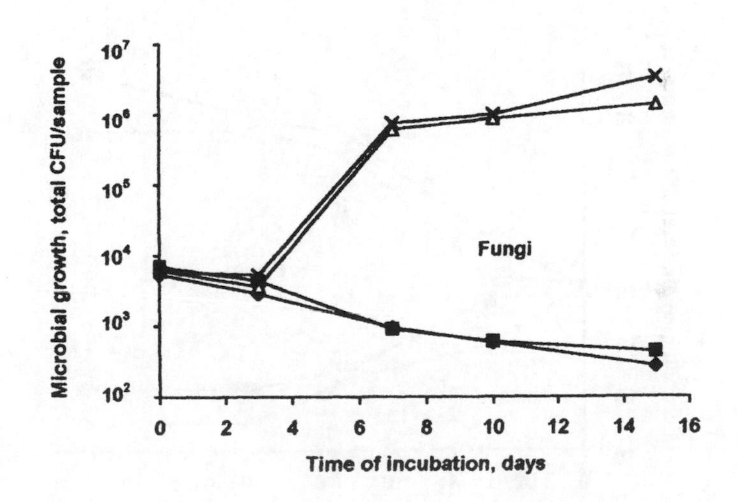

Figure 6. Growth of bacteria (top) and fungi (bottom) on heat treated fibroin. Fibroin threads were not heated (control) ◆-◆ or heat treated for different periods of time to give agings corresponding to 3 years ■-■, 6 years Δ-Δ, and 10 years **x-x** of storage.

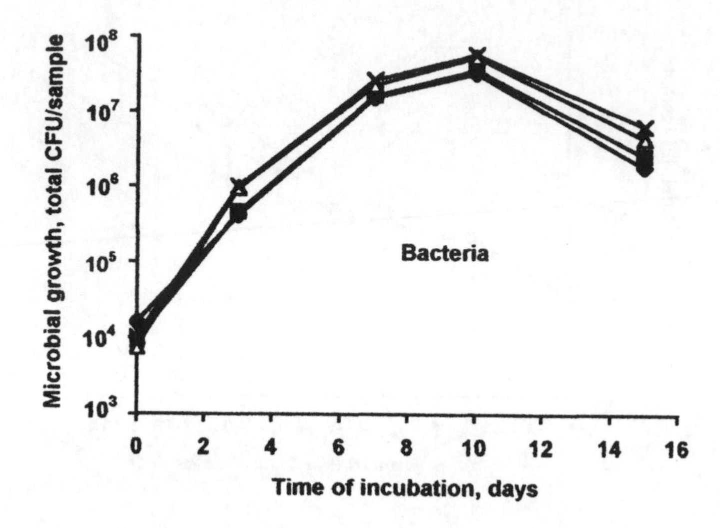

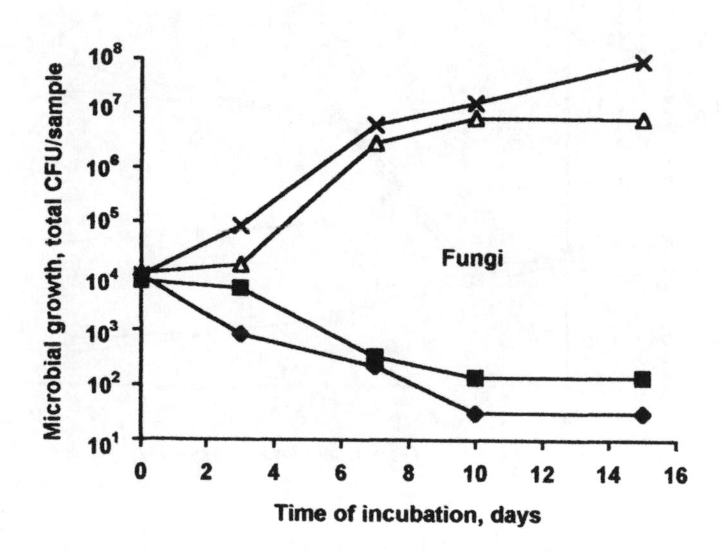

Figure 7. Growth of bacteria (top) and fungi (bottom) on fibroin exposed to a Xenon light. Fibroin threads were not illuminated (control) ◆-◆ or exposed to a Xenon lamp for 60 hours ■-■, 120 hours Δ-Δ and 193 hours **x-x**.

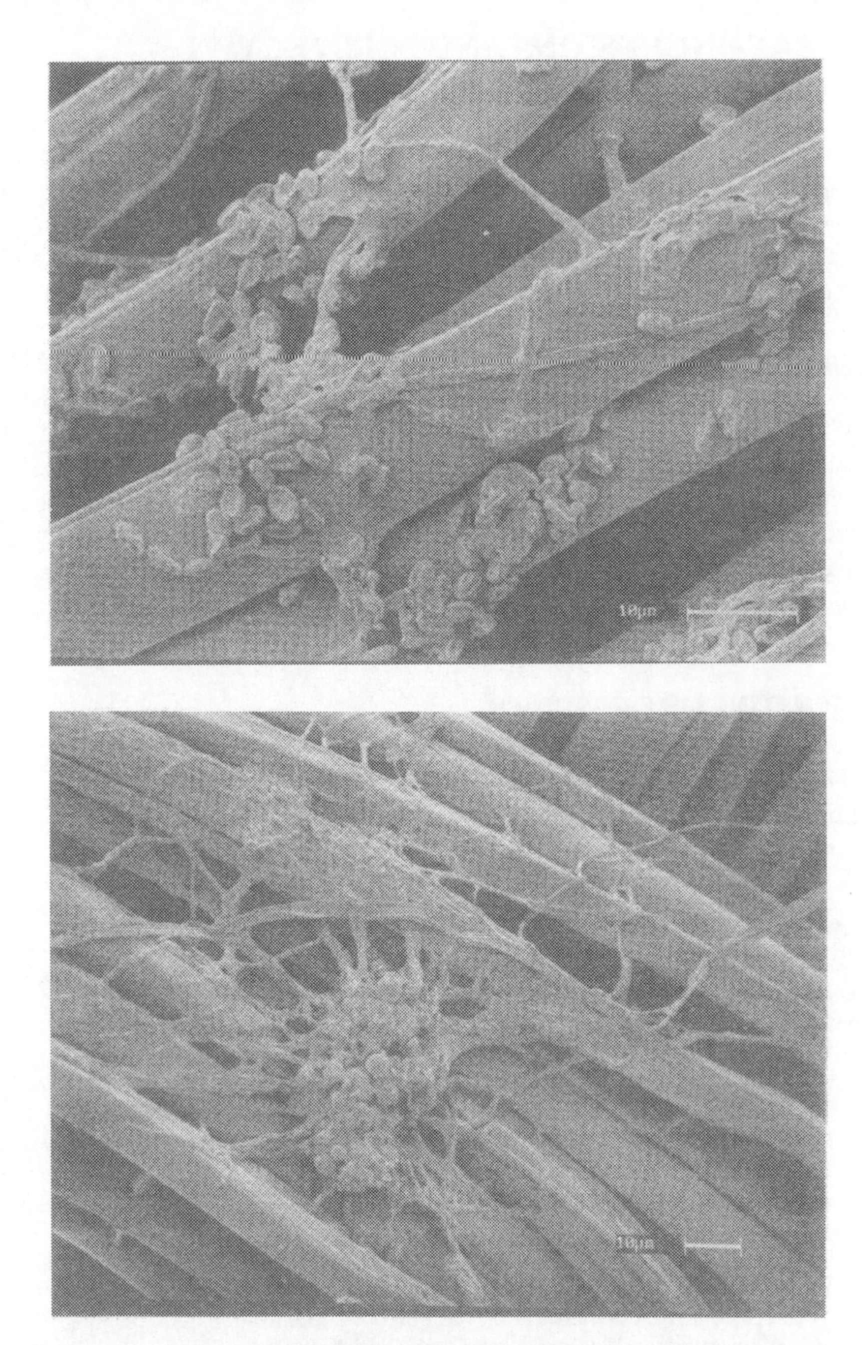

Figure 8. Fungal growth on fibroin aged in the laboratory by heating (top) or exposing to a Xenon lamp light (bottom). Spores and mycelium of *Cladosporium cladosporiodes* are evident on the heat-treated sample whereas spores and mycelium of this fungus and of *Aspergillus niger* are present on the sample exposed to light.

3. CONCLUSIONS AND PERSPECTIVES

In conclusion, it seems to us that laboratory models may allow to identify the microbial taxa defacing cultural heritage. Further, such models may permit to distinguish between the microorganisms responsible for the damage (we could call them the parasites) and those present in the microbial communities but not contributing to the defacement (the saprophytes). This type of experiment may also permit to monitor and evaluate the onset and the rate of microbial colonization, and the changes occurring in the microbial populations as a function of the substrate's composition and the environmental conditions. Finally, these models provide information essential for choosing the compounds to be used for disinfestation and suggest the most suitable materials to be employed for restoration and relining. If, for any reason, control of humidity, temperature, and light, as it occurs in museums, is not possible, then protection of objects of artistic or historical interest rests only on their intrinsic composition that should render such objects refractory to microbial colonization.

ACKNOWLEDGEMENTS

This work was supported by grants from Progetto Finalizzato Beni Culturali of the Italian National Research Council (C.N.R.).

REFERENCES

Bock, E. and W. Sand. 1993. The microbiology of masonry biodeterioration. J. Appl. Bacteriol. **74:** 503-514.

Kurupillai, R.V., S.P. Hersh and P.A. Tucker. 1989. Degradation of silk by heat and light. *In* Historic Textile and Paper Materials. Needles, H.L., Zeronian, S.H., Eds.; Advances in Chemistry **212:** 111-127. American Chemical Society, Washington, D.C.

O'Neill, T.B. 1986. Succession and interrelationships of microoganisms on painted surfaces. J. Coatings Technol. **58:** 51-56.

Ortega-Calvo, J. J., M. Hernandez-Marine and C. Saiz-Jimenez. 1991. Biodeterioration of building materials by cyanobacteria and algae. Int. Biodeterioration **28:** 165-185.

Reynolds, E.S. 1950. Pullularia as a cause of deterioration of paint and plastic surfaces in South Florida. Mycologia **42:** 432-448.

Seves, A.M., S. Sora and O. Ciferri. 1995. The microbial colonization of oil paintings. A laboratory investigation. 1st Congress on Science and Technology for the Safeguard on Cultural Heritage in the Mediterranean Basin 26 nov.-2 dec., Catania, p. 1315-1317.

Seves, A.M., S. Sora and O. Ciferri. 1996. The microbial colonization of oil paintings. A laboratory investigation. Int. Biodeterioration and Biodegradation **37:** 215-224.

Seves, A.M., M. Romanò, T. Maifreni, S. Sora and O. Ciferri. 1998. The microbial degradation of silk: a laboratory investigation. Int. Biodeterioration and Biodegradation **42:** 203-211.

Seves, A.M., S. Sora, G. Scicolone, G. Testa, A.M. Bonfatti, E. Rossi and A. Seves. 1999. Effect of thermal accelerated ageing on the structural properties of model canvas paintings. Science and Technology for Cultural Heritage. In press.

Wazny, J. and P. Rudniewski. 1972. The biodeterioration of binding materials used in artistic painting. Material und Organismen **7:** 81-92.

Winters, H., I.R. Isquith and M. Goll. 1975. A study of the ecological succession in biodeterioration of a vinyl acrylic paint film. *In* Developments in Industrial Microbiology. Vol. **17:** 167-171.

FUNGAL FOX SPOTS AND OTHERS

Nomenclature, SEM Identification of Causative Fungi, and Effects on Paper

Mary-Lou Esther Florian
Research Associate and Emerita, Royal British Columbia Museum, Victoria, BC V8V 1K5, Canada.

Key words: foxing, conidia, *Eurotium*

Abstract: Discoloured spots in 20 books dating between 1568 and 1902 were examined visually and by stereoscopic examination. Selected spots were examined by SEM and for EDX elemental analysis. The purpose was to determine if a nomenclature based on visual and stereoscopic non-destructive analysis was possible. It was possible to design a nomenclature based on surface appearance, shape, and colour. The following nomenclature is suggested: irregular fungal fox spot, circular fungal fox spot, irregular contemporary fungal spot, circular contemporary fungal spot, corroded iron spot, metal particulate spot, calcium particulate spot and protein spot. The commonest spots are "irregular fungal fox spots" and "corroded iron spots".
Viability tests on 114 irregular fungal fox spots from one book, dating 1854, showed that the fungal structures were not viable thus SEM analysis of 84 irregular fungal fox spots was undertaken to determine the distribution and to identify the causative species present. There were mainly two species which were identified by their conidia size, shape and ornamentation. The two species are probably *Eurotium spp* which are xerophylic species. They are the causative fungal species in irregular fungal fox spots. The two species were distributed randomly throughout the pages in the book suggesting contamination during the paper or book making process.

1. INTRODUCTION

This paper covers the characterization of different discoloured spots on paper in old books and suggests a simple nomenclature for the spots based on their visual appearance and cause. It also presents the results of an extensive examination of 85 fungal fox spots in one book undertaken to determine the variety of fungal species, their identification and the method and time of contamination. Information gleaned from this study on the effects on the paper material is also presented.

2. NOMENCLATURE OF COMMON SPOTS ON OLD PAPER

2.1 Initial comments

The term foxed spot, fox spot or foxing originally was used in reference to the rusty red spots which were similar to the red colour of the fox fur (Beckwith et al., 1940). There is extensive conservation literature that describes all aspects of foxing or fox spots on paper. This literature is thoroughly reviewed by Bertalan (1994). In this literature these terms are loosely used to include all discolouration.

In devising a nomenclature for the spots it is obvious that foxing or fox spot should be used when the spot is discoloured rusty red. Spots may have many different appearances and colour but only those that are rusty red should be called fox spots. Thus there is a need to have a nomenclature for the other spots.

The causes of and descriptions for some types of spots have been reviewed in the literature (Beckwith et al., 1940; Meynell and Newsam, 1978; Tang, 1978; Cain and Miller, 1984; Gallo and Hey, 1988; Daniels and Meeks, 1988; Gallo and Pasquariello, 1989; Linterink et al., 1991; Arai, 1993; Cain, 1993; Strzelczyk and Pronobis-Bobowska, 1993; Florian, 1996; Choisy, 1997). Iron, fungi and moisture condensation processes are suggested as possible causes of foxing. Several papers suggest names or a nomenclature for the different spots but there is no consistency in terms.

Cain (1993), and Cain and Miller (1984) have named various spots: **bullseyes** with dark centre and concentric rings; **snowflakes** with light brown areas having scalloped edges and without a central spot; **offprints;** and **shadows**. The latter two are related to the print on a page. The bullseye spots are associated with a central iron fragment, and most of the snowflakes have a small increase of iron concentration, within 10% over the background. In this latter class fungal hyphae were noted in a few spots and the spots showed variable responses to ultraviolet absorbency.

Strzelczyk and Pronobis-Bobowska (1993) proposed names for four main types of foxing spots; **eyes, clouds, freckles,** and **stars**, based on visual appearance.

Choisy et al. (1997) have categorized fox spots using the Fourier Transmission Infra Red (FTIR) spectrum. They felt from their examination of fox spots that the appearance of fox spots may be a bad key for taxonomy and for this reason tested FTIR as a non-destructive method providing some chemical information. They found three categories with subcategories which are described according to the chemical groups: non-conjugated ketones which may come from the degraded cellulose; unsaturated compounds (+C, C+N, C+O); and conjugated systems of C+C and C+O or C+N linked to sugars. The analysis of the spot includes the coloured component as well as breakdown products of the paper fibres. They state that the FTIR analysis is difficult to interpret and propose that extraction and chemical analysis of the purified solution is needed to provide the answer, which they are presently researching.

None of the above studies examined the spots with light microscopy to determine the location of the discolouration, the presence of fungal structures or the condition of the paper fibres in the spots. Florian (1996) describes in detail one type of spot, the rusty red, irregular or diffuse shaped spots of fungal origin, which she described as an **irregular fungal fox spot**. Each spot was examined visually, and with stereoscopic and transmitted light microscopy, Scanning Electron Microscopy (SEM) and energy dispersive xray spectroscopy (EDX) elemental analyses to determine their cause, the location of the stain and the condition of the paper fibres.

DNA studies are also being suggested as a method of characterizing fungal fox spots (Eveleigh et al., 1999).

The present author felt that a nomenclature for the spots based on a visual or stereoscopic microscope examination, which does not require expensive equipment and is non-destructive, seemed necessary and logical for paper conservators.

2.2 Results and discussion

All types of spots on paper from twenty books dating between 1568 and 1902 were examined. The books were printed in England, Canada, United States and Holland and came from libraries in England, South Africa, Brazil, Canada and USA. The books used had no heritage value and were experimental samples thus destructive analysis was possible, but the final nomenclature is based on non-destructive visual and stereoscopic examination. One sample of contemporary fungal damaged bond paper was used for comparison. Tide margins, obvious spots from soiling or splashes, were not included in this study. To thoroughly understand the cause of the spots and suggest a nomenclature based on their visual and stereoscopic appearance, selected spots were examined by stereoscopic and light microscopy, SEM, Hitachi (S-570); and elemental analysis by EDX, Kevex 8000, using 350 magnification, 500 count rate and 200 minutes elapsed time. Details of the methods are described in Florian (1996). In that reference only fungal fox spots are described.

To make it possible to examine one spot under all methods, a circle of the paper with the spot in the centre were cut out. For SEM and EDX analyses a pie shaped section was cut from this circle with the centre of the spot at the point of the pie piece. Having the spot at the point of the sample made it easy to locate the precise centre of the spot under the SEM. The remaining part of the spot, with the pie piece removed, was used for making permanent and temporary slides for light microscopy. The outermost margin away from the point was used as the control.

Ultraviolet (UV) fluorescence was not used in this project because of discrepancies in fluorescence of foxed spots reported in the literature. Cain and Millar (1982) found that of 360 rusty red/brown fox spots only one quarter showed fluorescence, they also found many fluorescent spots which could not be seen visually. Gallo and Pasquariello (1989) found the same inconsistency in fluorescence of rusty brown spots. Choisy et al. (1997) in their fluorescence examination found brown and yellow brown fluorescent spots. Florian and Manning (1999) reported that on UV examination of over a hundred irregular fungal fox spots in one book, published in 1854, all showed yellow fluorescence. Meynell and Newsam (1978), Cain and Millar (1982), and Choisy et al. (1997) theorize that new spots, which have not developed the foxed colour, show only fluorescence and as they become older the fox colour develops slowly to reach full colour and then it loses its fluorescence. Meynell and Newsam (1979) reports that the youngest rusty red fox spots he has observed were 23 years old. To understand the discrepancies is a major research project.

The following are the descriptions of the spots as seen by visual observation (VO), stereoscopic microscopy (SO), light microscopy (MO)and SEM, and EDX analysis.

Irregular fungal fox spots. Common on mid-19th century rag paper, contamination occurred during paper or book making.

VO - Irregular shaped, rusty red, uniformly discoloured, randomly present on all parts of a page and on all pages in a book, size varies from just visible to large areas on a page, usually migrated to adjacent pages.

SO - No alteration to surface of paper (Fig. 1), surface fungal structures rarely observable.

LM - A few hyphae, conidia and rarely conidiophores, protein material associated with area of fungal structures (Fig. 2) cell wall of paper fibres under the fungal structures are unevenly darkened.

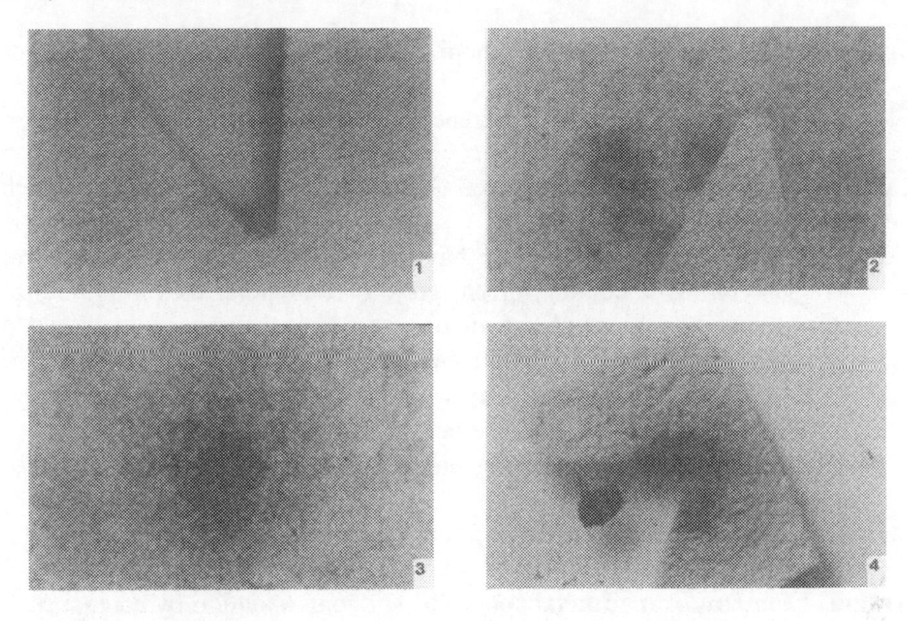

Figure 1. Surface of irregular fungal fox spot under stereoscopic microscope (SO). There is no change in the surface except for the rusty red colour.
Figure 2. Same spot as in Fig. 1 stained to show the protein associated with presence of fungi.
Figure 3. Corroded iron spot with central iron particle and impregnated adjacent fibres under SO.
Figure 4. Corroded iron spot with central spot dislodged (SO).

SEM - A few isolated fungal structure.
EDX - No iron or any elemental change in the paper under the spot.

Circular fungal fox spots. Occasionally on old paper or water damaged old paper, due to airborne contamination during use of book or paper.

VO - Circular shaped rusty red, uniformly discoloured, or may be concentrated in the centre, occurs most commonly towards margins of the page, and common on water damaged paper, size varies from pin point to larger spots dependent on growth, usually migrated to adjacent pages.

SO - Usually no alteration to the paper surface, but usually appears dusty due to presence of fungal colony growth, rarely cellulose digesting species may cause degradation of paper fibres.

MO - Many fungal structures, cell wall of paper fibres under fungal structures unevenly darkened, rarely cellulose digesting species may cause degradation of paper fibres.

SEM - Hyphae, conidia and conidiophores present, rarely ascocarps or cleistothecium present.

EDX - No iron or any elemental difference in the paper under the spot.

Circular contemporary fungal spots. Recent airborne contamination which show the colour of the conidia.

VO - Circular fungal colony, may have white fluffy margin and central region colour of conidia, green, grey, black, white, etc., vary in size depending on amount of growth, on one side of the page.

SO - Mass of hyphae and coloured conidia.

MO - Profuse fungal structures, paper under spot may be discoloured by pigments (red, yellow, etc.) secreted by fungus.

SEM - Profuse hyphae, conidiophores and conidia, ascocarps and ascospores may also be present.

EDX - No elemental change in paper under spot.

Irregular contemporary fungal spots. Recent contamination by contact with contaminated materials or water damage, they still show the conidia colour.

VO - Variable size diffuse, irregular, shaped areas or on the complete surface, may have wispy or profuse surface growth, may have colour of conidia, green, grey, black, white, etc., only on one side of the page, no alteration to paper surface.

SO - Wispy or profuse fungal growth the colour of conidia, paper under spot may be discoloured by pigments (red, yellow, etc.) secreted by fungus.

MO - Little or profuse fungal structures, ascocarps and ascospores may also be present.

SEM - Few or many hyphae, conidia and conidiophores, ascocarps and ascospores may also be present.

EDX - No elemental change in paper under spot.

Particulate spots. Spots caused by contaminating particles deposited during paper making or treatment, rare, rarely involve the adjacent fibres, usually less the 1mm. Only on one side of the page.

Corroded iron spots. Iron particle deposited during paper making process, commonly observed on alum treated paper because the treatment causes corrosion of iron.

VO - Circular spot, black to brown central spot and lighter grey margin, size varies from pin point to 2mm, no migration to adjacent pages, usually only on one side of the page unless deterioration through the page occurs.

SO - Black to brown central spot raised above paper level, paper fibres are impregnated with brown corrosion product, dark in centre and lighter towards the outer margin giving it a halo look (Fig. 3), the black central particle is easily separated from the paper ((Fig. 4) the rusty red colour of the corroded iron is masked by the discoloured alum treatment.

LM - No fungal structures, the paper fibres are fractured and brittle.

SEM - Central particle of iron oxide plaques, paper fibres are fractured.

EDX - Iron is present in particle and adjacent impregnated fibres.

Metal particulate spots. Rare

VO - (black, green, silver) spot, discolouration is usually restricted to the particle, usually less than a millimetre, may be slightly raised, near circular, on one side of the page.

SO - Central particle is embedded in paper fibres, corrosion of metal may cause slight eruptions on surface, spot is slightly raised, adjacent fibres rarely may have colour of corrosion products, i.e. green copper chloride, which appear as a halo around the particle.

MO - No fungal structures, particle embedded between fibres, adjacent fibres may or may not have colour of particle.

SEM - Particle embedded in paper fibres. Usually no change to paper fibres unless corrosion has occurred.

EDX - High zinc or copper peak or both for brass.

Calcium particulate spots.

VO - Oblong, white to cream spot, whiter than the paper, 1-2mm in length, no migration to adjacent pages.

SO - Central particle is embedded in paper fibres, spot is slightly raised, paper fibres in the spot are whiter than adjacent fibres, oblong shape suggests squashed by rolling mechanism.

MO - No fungal structures, particle crushed between fibres.

SEM - Crushed particle embedded in paper fibres, no alteration to paper fibres.

EDX - High calcium peak.

Protein spots.

VO - Light grey to brown, may appear shiny, sunken, 1-2mm., only on one side of a page.

SO - Glob of sunken material on top of paper with dull to shiny surface.

MO - Fungal structures may be attached to glob and extend out of the margin, hairs and fibres may be embedded in glob.

SEM - Glob of protein with fibres, hairs embedded in it.

EDX - No elemental change.

It was found that a nomenclature based on the visual and stereoscopic examination of the spots was possible. The result showed that there were basically two common types of spots and only a few other rare types. This is much better than the over 50 spots tabulated in 14 tables by Dalton and Wiltshire (1949) who reported over 50 causes of spots and specks on paper, from lipstick through all the paper making chemicals and just plain dirt. The two types are **irregular shaped fungal fox spots** caused by fungi on rag paper and **corroded iron spots** caused by corroded iron particles on alum treated paper. These two types of spots were present in varying numbers on all pages throughout a book unless the book contained paper of different origin.

The irregular fungal fox spots do not have iron amounts above the background paper (Beckwith et al., 1940; Arai, 1987; Florian, 1996). Beckwith et al. (1940) reported that there was no iron oxide or iron metal in the fungal spot.

3. IDENTIFICATION AND DISTRIBUTION OF THE FUNGAL SPECIES IN THE IRREGULAR FUNGAL FOX SPOTS IN ONE BOOK

3.1 Introduction

The literature review on foxing (Bertalan, 1994) shows that many fungal species have been identified from cultures that developed from samples taken from fox spots. Hundreds of species have been cultured and identified. The species that were identified are common airborne cosmopolitan fungi and not substrate specific. The majority are from the conidial fungi group mainly *Aspergillus*, *Penicillium* and Fungi Imperfecti (*Deuteromycetes*), which make up the major group of surface fungi found on all mouldy heritage materials, food, grain, paper, etc.

Florian (1997) has reviewed the literature on conidia survivability which suggests that conidia rarely remain viable for longer than 20 years, even under special storage conditions. Thus the viable conidia that were cultured must have been from recent contamination or young spots.

The viability of the irregular fungal fox spots in 114 spots in the 1854 book were tested and were non-viable. Spots on a page were outlined with graphite pencil, the page was then cut to fit into a petri plate and was placed on sterile culture media (Florian and Manning, 1999). There were a few fungal colonies which formed but they were not in the marked spot, thus were experimental or recent paper contamination. Because the fungal structures in the spots were not viable it was decided to test if the ornamentation of the conidia, seen under SEM, could be used to identify the fungal species. This would insure that the species identified would be the causative fungi. Analysis of the distribution of the fungal species on a page and the pages throughout the book and their relationship to the print was undertaken in hopes of determining the method and time of contamination.

3.2 Results and discussion

3.2.1 Overall description of the fungal structures and distribution of species

The book used for experimentation was "Lives of Brothers Humboldt, Alexander and William", by T.D. Bauer, 1854, published by Harper and Brothers Publishers in N.Y. It was unique in that the first 96 pages were alum treated and the remaining 259 pages were not treated. The paper was rag paper. Eighty-four **irregular fungal fox spots** (Fig. 1 and 2) from 63 untreated pages were examined by SEM. Irregular fungal fox spots and non-foxed areas for controls were cut from the pages using a razor blade or scalpel and, with tweezers, were placed on two-sided sticky tape already stuck to the aluminium SEM stubs. The fungal structures in the spots had air dried in the book thus no pretreatment was considered necessary. They were coated with gold/palladium using a Technics Hummer V sputtercoater. The coated samples were examined using a JEOL JAM-35C scanning electron microscope (SEM). Micrographs were taken on Polaroid positive/negative 665 film using 10KV to reduce surface charging and edge effects and at 15KV when higher resolution was required.

The fungal fox spots contained only isolated fragments of fungal structures, no colony growth was observed . All the spots contained a few conidia and hyphae, but only a very few spots contained fragments of conidiophores.

The results showed basically conidia of two different ornamentation, Type 1 (T1) (Fig. 5) and Type 2 (T1) (Fig. 6)and a small group of conidia with different ornamentation than the common species, Type3 (T3) (Florian and Manning, 1999). The T1 and T2 conidia were easily separated because of their different shape, size and ornamentation. A total of 84 spots were examined, T1 was found in 55 %, T2 in 32% and T3 in 13%. Only 5 spots contained more than one species.

The two main species were randomly distributed on the pages throughout the book. This means that the pages were contaminated by something which was itself contaminated with the fungal species. Type 1 when in the text region was on top of the printed text and T2 when in the text region was under the text. This suggests that the T1 species contamination of the paper occurred during or shortly after the printing process and the T2 species prior to printing.

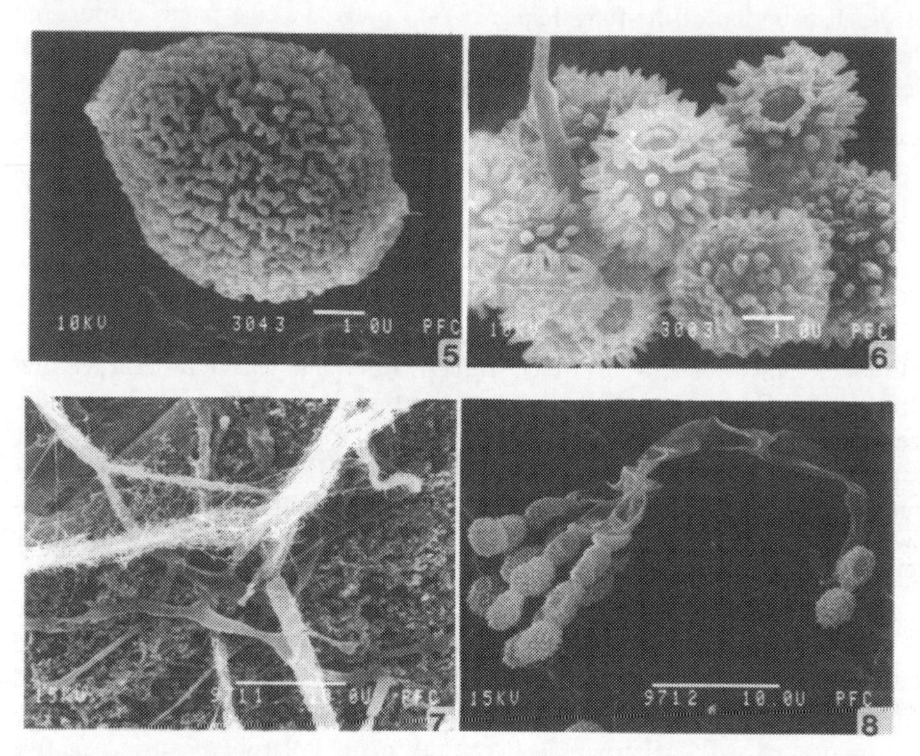

Figure 5. Type 1 conidia, possible ***Eurotium amstelodami*** Mangia. Elliptical with lobate-reticulate ornamentation.

Figure 6. Type 2 conidia, possible ***Eurotium repens*** de Bary or ***E. rubrum***. Globose with aculeate ornamentation.

Figure 7. Type 1 surface hyphae is covered with hairy mycofibrils and Type 2 smooth hyphae in the paper shows bacterial damage.

Figure 8. A fragment of a Type 2 conidiophore. The conidia show the variation in ornamentation due to stages of development.

The T1 and T2 species had different hyphal characteristics which suggested that they came from two different environments, thus two different sources. T1 had hyphae with profuse mycofibrils (fluffy covering) (Fig.7) common on aerial hyphae and T2 had hyphae with bacterial lytic holes (Fig. 7) suggesting that they were from a wet environment. T1 hyphae were occasionally seen between the paper fibres but T1 hyphae were always in fragments on the surface of the paper.

The few conidiophores (Fig. 8) were associated with T2 conidia, but only one conidiophore was found with T1 conidia. The conidiophores assist in determining the genus of the fungus, and the conidia ornamentation the species.

Another book, published in 1785, was analysed for comparison and showed a similar random distribution of two fungal species. The species are different than those in the first book analysed. This suggests that this type of analysis may be useful in authenticity studies.

During the examination, silk threads, faecal pellets, fractured egg shells, and a mite-like animal were observed which suggests a complex ecosystem.

3.2.2 Identification of genus

In the *Aspergillus* genus the conidia develop on a thin erect structure called the conidiophore. The conidiophore has a rounded tip which is covered with finger-like conidiogenous (producing conidia) cells. The conidia are borne on the tip of each terminal conidiogenous cell in long chains, giving the head a hairy look.

In the *Penicillium* genus conidiophores do not have a globose head. The conidiophore branches like a flower inflorescence which has many small branches. The condiogenous cells and their chains of conidia develop at the tips of the joints of the branches at different levels.

These characteristics are used to determine the genus.

The conidiophore fragments (Fig. 8) from T1 and T2 species were examined by three mycologists and myself, the tally was two for *Penicillium* and two for *Aspergillus*. The problem in identifying the conidiophores is because of the collapse and shrinkage due to dehydration. In SEM preparation of fresh, living fungal material, many methods have been tested to determine how to obtain the least distorted shapes.

Staugaard et al. (1990) reported on the variation in the morphology of *Penicillium* and *Aspergillus* conidia, conidiophores and conidiogenous cells (stipes, metulae and phialides) caused by different SEM preparatory methods. The methods compared were: observation at low temperature of frozen hydrated material (cryofixation, cryoSEM, cryoscanning), chemical fixation and critical point drying (CPD). They showed that cryofixation and light microscopy of normally hydrated material gave the best result. The other methods showed shrinkage and deformation.

Jones et al. (1996) noted some differences in the morphology in the conidiophores when grown on different media, as well as from different SEM preparatory methods. They reported that cryofixed images give the more exact shape.

The samples used in this project had no SEM preparatory treatment because they were already air dried, thus the dehydrated conidiophores had already shrunk and distorted. With this in mind, and on thoughtful observation of the conidiophores, they resemble the *Aspergillus* genus more than *Penicillium*.

3.2.3 Identification of the species by conidia ornamentation

3.2.3.1 The methods of SEM influence the ornamentation

The identification of the species is usually done by observation of the characteristic of a growing culture, but it can be assisted by comparing the SEM of conidia ornamentation, size and shape with SEM illustrations of known species in the literature (Locci, 1972; Martinz et al., 1982; Ramirez, 1982; Torras-Calvo, 1982; Kozakiewicz, 1984; Kozakiewicz, 1989a; Kozakiewicz, 1989b).

The conidia has the same problems with shrinkage and distortion due to dehydration as the conidiophores. Staugaard et al. (1990) reported that some markings can not be seen on hydrated conidia but are present on naturally air dried conidia. Kozakiewicz (1983, 1989a) showed that there were conidial ornamentation changes during development terminating in the mature conidia which had a fixed and unchanging ornamentation for both *Aspergillus* and *Penicillium* species. Thus mature conidia must be used for reference. Bridge et al. (1986) found that on culturing single conidia strains there were differences in the conidia ornamentation and size probably due to mutations. Kozakiewicz (1983) reported that the conidia ornamentation pattern remained constant in species subjected to physiological and environmental stress.

The size of the conidia also varied with dehydration methods. Staugaard et al. (1990) state that the dimensions of the conidia are dependent on age and growth of the fungal material and to a great extent on the preparatory technique. They report dehydration shrinkage rates of 8-34% with an average of 20-30%. Different degrees of shrinkage occur with different species. The comparisons were based on differences with conidia in 80% lactic acid measured under light microscopy. Read et al. (1983) found that air-dried conidia probably showed most shrinkage while frozen hydrated conidia the least shrinkage. The haunting thought arises when using 145 year old conidia. They are the ancestors of the present day conidia, have mutations occurred in the species over the years.

Thus in using these published illustrations of known species for comparison, first the species being identified must be represented in the illustrations, the SEM preparatory methods should be the same as was used for the illustrations and mature conidia must be used.

The SEM illustrations of the ornamentation in the literature used for fungal species identification unfortunately is sparse and the methods of preparation vary. The excellent article on *Penicillium* species by Kozakiewicz (1989b) uses cryoscanning (CryoSEM) to prevent the artifacts associated with air-drying and critical-point drying techniques, but her paper (Kozakiewicz, 1989a) on *Aspergillus* species uses direct deposition of conidia which causes natural air drying.

Because the ornamentation is more obvious when the conidia is air dried naturally and it is the fastest and simplest method, I am surprised that this is not the method of choice. This would allow comparisons with non-viable conidia, that have naturally dried, found on contaminated artifacts as well as food materials, stored grain and in air samples.

3.2.3.2 The species and conidial ornamentation

The terminology for the ornamentation of conidia varies with author. Some terms used are smooth-walled, delicately roughened, warty, echinate, striate for *Penicillium* (Martinez et al., 1982). *Aspergillus* has nine categories (Kozakiewicz, 1989a), echinulate, aculeate, tuberculate, verrucose, lobate-reticulate, miro-tuberculate, micro-verrucose, striate, and smooth.

Dr. Zofia Kozakiewicz (1999) examined the conidia and conidiophores of the two species T1 and T2, and the following are her descriptions.

Type 1 (1) (Fig. 5): "This could possibly be identified as *Eurotium amstelodami* Mangin. Spore size in *E. amstelodami* is 4.5-5.0 x 3.5-4.0um. Which would fit. *E. amstelodami* spores are characterized by their subglobose to elliptical shape, bearing pronounced attachments at either end of the spore. In the SEM spores have a lobate-reticulate ornamentation, where the lobes are long flattened ridges formed in parallel lines."

Type 2 (Fig. 6): "This possibly could be *Eurotium repens* de Bary. Spore size in this species is from 4.5 to 7 or 8 μm, but mostly 5.0-6.5 μm. Spores are ovate to subglobose or globose in shape. Under the SEM spore ornamentation is aculeate. The type specimen for *E. repens* was isolated from badly dried herbarium material.

However, the species could possibly be *Eurotium rubrum* Konig, Spieckermann and Bremer. It is closely related to *E. repens*. Spores are ovate to subglobose or globose in shape, and 5-7.5 μm in diameter. Conidia have an aculeate ornamentation under the SEM."

She mentions that for verification, examination of living cultures would be necessary.

Eurotium is a species in an *Aspergillus* species group. The *Aspergillus* genus can be separated into 5 species groups (Kozakiewicz, 1989) which reflect natural association by similarities in colony appearance, micromorphology, and ecological data, *A. niger*, *A. glaucus*, *A. restrictus*, *A. fumigatus*, and *A. alutaceus*. *Eurotium* species are in the *A. glaucus* group.

Kozakiewicz (1999) points out that the *Eurotium* species are xerophilic. That is they are able to grow on products of low water activity (containing very little free water).

The significance of xerophilic fungi on paper is reviewed by Florian (1997). Xerophilic species have been isolated form contemporary contamination of paper. Arai (1987) isolated 25 strains, using the culture technique, from foxed area on hemp paper. The species were divided into two groups according to their water activity, 7 required high water activity and the remaining were able grow in both high and low water activity. Three of the latter group were *Eurotium herbarium* strains, and four *Aspergillus penicilloides spegazzini* strains. Nol et al. (1983) isolated 6 species of *Aspergillus*, and one each of *Penicillium* and *Gliocladium* species from blotches on contemporary postage stamps, all species showed some cellulase activity and 4 species were xerophilic. These two reports suggest that xerophilic fungi are contaminates on paper. Other fungal species are xerophylic. Pitt (1975, 1981) reports that from a list of 44 known xerophilic species, 12 were of *Aspergillus* genus, 14 *Penicillium* and 7 *Eurotium*. *Aspergillus* species have a lower minimum water activity requirement than the *Penicillium* species.

4. CONCLUSION

It is possible to identify the different types of discoloured spots on old paper by the use of visual and stereoscopic examination. The following nomenclature is suggested: irregular fungal fox spot, circular fungal fox spot, circular contemporary fungal spot, irregular contemporary fungal spot, corroded iron spot, metal particulate spot, calcium particulate spot and protein spot. The commonest spots were irregular fungal fox spots and corroded iron spots. The nomenclature for the spots is based on their surface appearance, colour, and shape. The term fox should be used for only those spots which have the rusty red colour of the fox fur. The old fungal spots are rusty red and contemporary fungal spots are the colour of their conidia. The shape of the spot suggests the method of contamination, circular spots occur from airborne conidia, and irregular diffuse spots from surface contamination from contaminated materials or during manufacturing. Particulate spots are caused by contaminating metal fragments in the paper and protein globs from the size.

Establishing a nomenclature is essential for future research. To make the research meaningful we must describe the type of spot, its age, cause and method of contamination.

The extensive SEM examination of 84 irregular fungal fox spots in one book published in 1854, showed that there were mainly two fungal species (T1 and T2) in the one book. Their distribution on the page and on the pages throughout the book showed a random distribution. This suggests contamination from manufacturing or contact with contaminated materials during paper or book making. T1 had hyphae mycofibrils suggesting aerial growth and T2 had hyphae with bacterial damage which suggests it came from a wet environment. T1 was observed on top of the print and T2 appeared below the print with some hyphae between the paper fibres.

The two species were easily recognized by the conidia distinct differences in size, shape and ornamentation.

The two species were identified by the characteristics of their hyphae, conidiophores and conidia. They are probably *Eurotium* species. Even though wc can not be absolutely positive about their identification, the results are significant because they are the actual causative species in the irregular fox fungal structures in this 1854 book.

This study showed that the fungal contamination occurred during the paper or bookmaking process and suggests two different times of contamination.

ACKNOWLEDGEMENTS

The research is a result of the research assistance of following: Mary Magar, Metals and Material Dept., University of British Columbia; Dr. Brenda Callan and Lesley Manning, Canadian Government Pacific Forest Centre Research Laboratory; Betty Walsh, Conservation Laboratory, Archives of British Columbia; Valerie Thorp, Conservation Services Section, Royal British Columbia Museum, Dr. Lynn Siegler, Devonian Institute, University of Alberta, Dr. Zofia Kozakiewicz-Lawrence, CABI Bioscience, Surrey, UK; and the financial assistance from the Canadian Council of Archivists and the Getty Conservation Institute. I would like to thank the following for providing the book samples, Barry Byers, Conservation, B.C. Archives, Victoria, B.C., Canada; Nancy Purinton, Conservation, Harpers Ferry, US Parks; Kasia Szeleynski, Paper Conservation Tate Gallery; and Jack Thompson of Thompson Conservation, Portland, Oregon.

REFERENCES

Arai, H. 1993. Relationship between fungi and brown spots found in various materials. *In* Biodeterioration of Cultural Property, Proceedings of the 2nd International Conference, Yokohama, Japan, p. 320-336.

Arai, H. 1987. Microbial studies on the conservation of paper and related properties. Science for Conservation **26**: 43-52.

Beckwith, T.D., Swanson, W.H. and Iiams, T.M. 1940. Deterioration of paper: the cause and effect of foxing. Univ. of Calif. (L.A.) Biological Sciences Publication **1**: 299-356

Bertalan, S. 1994. Foxing. In Paper Conservation Catalogue, 9th edition, AIC **13**: 1-39.
Bridge, P.D., Hawksworth, D.L., Kozakiewicz, Z., Onions, A.H.S. and Paterson, R.R.M. 1986. Morphological and biochemical variation in single isolates of *Penicillium*. Trans. Br. Mycological Society **87**: 389-396.
Cain, C.E. 1993. Foxing caused by inorganic factors. *In* Biodeterioration of Cultural Property, Proceedings of the 2nd International Conference, Yokohama, Japan, p. 278-291.
Cain, C.E. and Miller,B.A. 1984. Proposed classification of foxing. American Institute for Conservation, Book and Paper Group, 10th Annual Meeting Postprints p. 29-30.
Choisy, P., de la Chapelle, A., Thomas, D. and Legoy, M.D. 1997. Non invasive technique for the investigation of foxing on graphic art material. Restaurator **18**: 131-152.
Dalton, J.A. and Wiltshire, W.A. 1949. The systematic identification of spots and specks in paper. Paper-making and British Paper Trade Journal **117**: 173-179.
Daniels, V.D. and Meeks, N.D. 1988. Foxing caused by copper alloy inclusions in paper. Proceedings of Symposium **88**: 229-233.
Eveleigh, D., Sullivan, R. and White, J. 1999. The role of fungi in foxing (characterization using PCR and DNA studies). *In* 11th International Biodeterioration and Biodegradation Symposium, Arlington, VA. (in press).
Florian, M-L.E. 1996. The role of the conidia of fungi in fox spots. Studies in Conservation **41**: 65-75.
Florian, M-L. 1997. Heritage Eaters: Insects and Fungi in Heritage Collections, James & James, London UK.
Florian, M-L. E. and Manning, L. 1999. The ecology of the fungal fox spot in a book published in 1854. *In* ICOM-CC, Lyons (in press).
Gallo, F. and Pasquariello, G. 1989. Foxing: ipotesi sull'origine bologica. Bollettino dell'Istituto Centrale per la Patologia del Libro Alfonso Gallo **43**: 139-176.
Gallo, F. and Hey, M. 1986. Foxing - a new approach. The Paper Conservator **12**: 101-102.
Jones, D., Vaughan, D. and McHardy, W. J. 1996. A critical examination of SEM ultrastructure features in two *Penicillium thomii* isolates from soil. Mycological Research **100**: 223-228.
Kozakiewicz, Z. 1983. Taxonomic variation in the Aspergilli of stored products using scanning electron microscopy and electrophesis. Unpublished PhD thesis of the University of London.
Kozakiewicz, Z. 1984. SEM taxonomy of *Aspergillus*. Royal Microscopical Society **1**: 216.
Kozakiewicz, Z. 1989a. *Aspergillus* species on stored products. Mycological Papers No. 161, C.A.B. International Mycological Institute, Surrey, UK. 1.
Kozakiewicz, Z. 1989b. Ornamentation types of conidia and conidiogenous structures in fasciculate *Penicillium* species using scanning electron microscopy. The Linnean Society of London p. 273-293.
Kozakiewicz-Lawrence, Z., PhD, Mycologist, Personal communications. 1999. CABI Bioscience, UK Centre, Egham, Bakeham Lane, Egham, Surrey ,UK.
Linterink, F., Parek, H.J. and Smit, W. 1991. Foxing stains and discolouration of leaf margins and paper surrounding printing ink: elements of a complex phenomenon in books. The Paper Conservator; Journal of the Institute of Paper Conservation **15**: 45-52.
Locci, R. 1972. Scanning electron microscopy of ascosporic Aspergilli. Rivista di Patologia Vegetale, Pavia **8**: 1-172 Supplemento V.
Martinez, A.T., Calvo, M.A. and Ramirez, C. 1982. Scanning electron microscopy of *Penicillium* conidia. Antonie van Leeuwenhoek **48**: 245-255.
Meynell, G.G. and. Newsam, R.J. 1978. Foxing, a fungal infection of paper. Nature **274**: 466-468.

Nol, L., Henis, Y. and Kenneth, R.G. 1983. Biological factors of foxing in postage stamp paper. International Biodeterioration Bulletin **19**: 19-25.

Pitt, J.I. 1975. Xerophylic fungi and spoilage of food of plant origin. *In* Water Relations of Food, R.B. Duckworth (ed.), Academic Press, London, p 273-307.

Pitt, J.I. 1981. Food spoilage and biodeterioration. *In* Biology of Conidial Fungi, G.T. Cole and B. Kendrick (eds.), Academic Press, NY, **2**: 111-142.

Ramirez, C. 1982. Manual and Atlas of the Penicillia. Elservier Biomedical Press, Amsterdam.

Read, N.D., Porter, R and Beckett, A. 1983. A comparison of preparative techniques for examination of the external morphology of fungal material with the scanning electron microscope. Canad. J. Bot. **61** :2059-2078.

Staugaard, P., Samson, R.A. and van der Horst, M.I. 1990. Variation in *Penicillium* and *Aspergillus* conidia in relation to preparation techniques for scanning electron and light microscopy. *In* Modern Concepts in *Penicillium* and *Aspergillus* Classification R.A. Samson and J.I. Pitt (eds.), Plenum Press, NY.

Strzelczyk, A. and Pronobis-Bobowska, M. 1993. Characteristics of foxing stains on paper historic objects. *In* Scientific Bases of Conservation and Restoration of Works of Art and Cultural Property Items, Uniwersytet Mikolaja Kopernika w Toruniu, Torun, Poland, p. 327-333.

Tang, L. 1978. Determination of iron and copper content in 18th and 19th century books by flameless atomic absorption spectroscopy, J. Am. Inst. Conservat. **17**: 19-32.

Torras- Calvo, M.A. , Guarro, J. and Torras, M. 1982. Surface characteristics of *Penicillia* conidia an electron microscopical investigation. Mikroskopie **39**: 85-89.

POLYMERS AND RESINS AS FOOD FOR MICROBES

Robert J. Koestler
Sherman Fairchild Center for Objects Conservation, The Metropolitan Museum of Art, 1000 Fifth Avenue, New York, NY 10028-0198, USA

Key words: biodeterioration, polymers, resins

Abstract: Preserving art objects, especially outdoor stone ones, from the ravages of man and time often entails application of a polymeric material. Unfortunately, some of these products may support or actually encourage microbial growth and cause unintentional problems. This paper summarizes studies that developed a screening test to quantify the effects of fungal activity on a variety of consolidants and resins, and FTIR-fungal studies on silicone-based resins. One of the studies was a five-week assessment of the ability of a mixture of six fungal species to use sixteen different polymer formulations. A quantitative ranking scale was developed for the products using a combination of organism growth, sporulation, weight change, and chemical change data. These data were applied in Duncan's Multiple Range test and Student-Newman-Keuls' test to statistically rank a series of conservation materials for their susceptibility to fungal attack. The products tested included three acrylics, five polyvinyls, five silicone-based polymers, one polyimide, and two natural resins. Damage to the products ranged from complete degradation to very little change. Some of the products that are degraded the most are quite prevalent in field use today (i.e., Conservare H40, Conservare OH, and Acryloid F-10). Fungal degradation may be related to the abundance of silanol molecules (Si-OH) for silicone-based products, C-H bonds for carbon-containing products, or the presence of 'inert' material used as bulking agents.

1. INTRODUCTION

Physical breakdown of materials is a subject that has long been studied with the aim of developing more durable materials and preserving art objects. Stone, one of the more durable materials, does nonetheless deteriorate. Deterioration of stone, whether in a man-made monument or in its natural form, is a constant and natural phenomenon, and can occur as a result of many causes, including those that are physical, chemical, mechanical, and biological. Recently researchers have begun to investigate how biological phenomena--at times in synergy with physical, chemical, and mechanical effects--can exacerbate deterioration of historically or culturally important stone materials (Koestler et al., 1994, 1997). Not only has the biodeterioration of stone become a topic of interest to the art conservation community, but also the susceptibility to biodeterioration of preservative materials applied to art objects has begun to be investigated (Koestler and Santoro, 1986, 1988; Leznicka et al., 1991; May et al., 1993; Nugari and Priori, 1985; Salvadori and Nugari, 1988). Preserving art objects, especially outdoor stone ones, from the ravages of time may entail application of one or more polymers or resins to the object. The material applied may be used as a waterproofing agent, as a (reinforcing) binding matrix to help hold the object together, or as a protective surface coating. Unfortunately, some of the products used may do more than expected and/or desired. They may act as a food source for and encourage the growth of microbes that inhabit the surface or interior of an object and lead to deterioration of that object.

There are many types of preservative treatments. The main types used today fall into four categories: silicone-based chemicals, which include silicone resins, alkoxy silanes, and silicate esters; synthetic organic polymers, which include acrylics, epoxies, vinyl polymers, and polyurethanes; inorganic materials, which include the alkali earth hydroxides such as barium hydroxide and various siliceous compounds such as silico-fluorides (many include the silicate esters within this group rather than with the silane-based treatments); and waxes and natural resins.

Assessing the resistance of a protective polymer or resin to microbiological attack is typically a lengthy process necessitating long-term field trials. Laboratory studies attempt to predict field trial results in a reduced time frame and have some advantages over field trials since micro-scale changes in the materials are more easily noted and studied in the laboratory. Often large-scale testing of materials can be avoided by application of a short-term screening test to narrow the field to those materials that are potentially the least susceptible to microbial attack. The screening study described in Part 1 assayed 16 polymers and resins for their resistance to fungal attack during a five-week screening period.

In addition to the screening study, seven polymers and resins were subjected to one or two months of fungal growth and then analyzed by Fourier transform infrared spectroscopy in an effort to quantify chemical bond changes that occurred as a result of biological action. This is summarized in Part 2.

The main source of biological contamination of stone is the surrounding soil, which usually contains large numbers and many different types of bacteria, fungi, and algae (Strzelczyk, 1981). These can contaminate the stone shortly after quarrying, or they may be within the stone before quarry extraction. Stone can also be contaminated by windblown detritus or by rising groundwater infiltration. The effect of microbial attack is not necessarily confined to the surface of the stone: Many studies have shown that microorganisms and organic matter may penetrate stone from depths of a few centimeters to almost a meter, especially, in cases of porous limestones or sandstones (Koestler et al., 1985; Krumbein, 1972). The organisms found within and colonizing the surfaces of rock and stone are essentially the same as normal soil microflora, both heterotrophic and chemotrophic organisms (Webley et al., 1963). The various microorganisms constitute a complex ecological intercommunity capable of carrying on the normal processes of stone weathering and soil formation.

Many bacterial heterotrophic species have been isolated from stone. Stone in the open air is likely to be covered and infiltrated with dirt and organic matter from rain, groundwater and airborne sources, and animal sources such as pigeon excrement, which has been seen to be a rich source of nutrients and encourages deterioration of marble statues (Bassi and Chiatante, 1976). These have been shown to contribute to the dissolution of siliceous and calcareous stones by the production of various organic acids, mainly 2-ketogluconic acid (Henderson and Duff, 1963). Bacteria isolated from calcareous sandstone monuments have been shown to have the ability to cause severe, rapid weight loss in sandstone by attack of the calcareous matrix of the stone with organic acids (Lewis et al., 1987). A variety of fungi have also been isolated from and associated with decaying stone (Bassi and Chiatante, 1975; Koestler et al., 1985); the fungi are generally assumed to attack stone by the excretion of organic acids, such as oxalic, citric, and fumaric acid.

The majority of previous studies on consolidants have focused on aspects of chemical or mechanical deterioration--not on biological deterioration. On a macro scale, it is of use to establish the relative durability of selected polymers and resins, in addition to understanding the mechanisms of deterioration. The screening study was designed as a way of measuring on a macro scale the relative biodurability of polymers and resins and thus providing the conservator with a fuller understanding of their strengths and weaknesses, in this case knowing the relative susceptibility of these polymers or resins to fungal deterioration.

Clearly the physical and chemical state of the active agent (e.g., consolidant, catalyst or additive) is important in assessment of microbiological attack (Hueck et al., 1968). After the active inhibitory agents fade, the protection value of the film as a barrier covering the material to which it was applied continues to be important. Some assessment of effects of washing and leaching of potential inhibitory components as related to weight loss were noted and are presented in the full study (Koestler and Santoro, 1988). Therefore, the stability of the coating over time is an important factor in evaluating its efficacy in preventing biological attack. Another important consideration is the permeability to moisture and gases; this will have a marked effect on microbial growth.

2. PART 1, THE SCREENING STUDY

For the 5-week screening study, materials selected were polymers and resins currently used in conservation or of research interest. The methodology of the biodeterioration testing has been reported in full in Santoro and Koestler (1991) and is summarized below. Using short-term laboratory exposures under conditions of high relative humidity, 16 samples of polymers and resins were evaluated for their ability to support fungal growth. Growth of the organisms was ascertained by assessing macroscopic, microscopic, and physico-chemical changes of these materials over the 5-week testing period. Based upon their sensitivity to fungal deterioration, the polymers and resins tested were ranked in order of least to most susceptible to biological attack.

Briefly, multiple samples of each consolidant, prepared as a thin film on large glass slides, were inoculated concurrently with two different concentrations (104 and 108 spores/ml) of fungal spore mixtures. Fungi utilized were: 2 *Penicillium* spp; *Fusarium* sp.; *Cladosporium* sp.; and 2 *Aspergillus* spp. Inoculated samples were grown in a temperature- and humidity-controlled chamber for 5 weeks and assayed weekly using grid matrix and random field procedures for signs of visible growths of fungi, for weight loss after fungal action, and for chemical changes to the consolidant after fungal exposure. Details of sample preparation and exposure conditions can be found in Santoro and Koestler (1991).

The polymers and resins tested are listed in the following table.

Polymers and resins tested in 5-week screening study

Product name	Manufacturer	Additives or contaminants
Acrylics		
Acryloid B-72	Rohm and Haas	Information not available (2)
Acryloid F-10	Rohm and Haas	Information not available (2)
Rhoplex AC-234	Rohm and Haas	Formaldehyde <0.08% (1) Ammonia <0.3%. (1) Probably soaps as emulsifiers
Polyvinyls		
Bakelite-AYAA	Union Carbide	No additives (2)
Bakelite-AYAC	Union Carbide	Acetic acid <0.05% (1)
Bakelite-AYAF	Union Carbide	No additives (2)
Bakelite-AYAT	Union Carbide	No additives (2)
Mowital B-20-H	Amer. Hoeschst	No additives; may be some byproducts from the polymerization process such as PVAC, PVOH, butyraldehyde (2)
Silicones		
Conservare H	ProSoCo	Dibutyltin dilaurate = catalyst (2)
Conservare H40	ProSoCo	Dibutyltin dilaurate = catalyst (2)
Conservare OH	ProSoCo	Dibutyltin dilaurate = catalyst (2)
Silicone 1048	General Electric	100% silicone resin (2)
Tegovakon V	Goldschmidt	Solvent + alcohol (2) Tin catalyst (2) Silicic acid ester = binder replacement
Polyimide		
Imron 192S	Dupont	HCl or triethylamine as an initiator (3) Activator 192S has no additives (2)
Natural resins		
Dammar	AF Suter	Natural resin
Shellac	AF Suter	Natural resin

(1) Manufacturer's literature and/or safety data sheet
(2) Phone call to manufacturer
(3) Jim Druzik, The Getty Conservation Institute, Los Angeles, CA

2.1 Organism growth assessment

Over the 35-day exposure interval, fungal growth was mixed, with Imron 192S, Dammar, Conservare H40, Mowital B-20-H, Bakelite-AYAF, and Bakelite-AYAT showing consistently high fungal growth over the testing interval.

Conversely, G.E. Silicone 1048, and Rhoplex AC-234 showed no fungal growth or minimal growth over the interval. Bakelite-AYAC showed little growth over 2 weeks at which time the exposed strips broke up--this was not seen in the controls and was therefore assumed to be related to fungal attack.

Indeed AYAC is not noted to be resistant to at least bacterial attack (product literature, Union Carbide Corp., USA).

In order to combine these polymers and resins into statistically significant groupings ($\alpha \geq 0.05$), a Wilcoxon's two-sample test was performed from the individual data points. Little consistency in the pattern of growth for polymers or resins within a similar group (e.g., polyvinyl acetates) was noted over the study (see Koestler and Santoro, 1988 for detailed discussion).

Comparisons of the random field procedure (RFP) to both visual and optical viewing techniques revealed that the RFP proved to be more precise than the visual scoring procedure and provided numerical assessment in a statistically valid manner, especially when compared to higher magnification viewing (Koestler et al., 1988). In many cases visual scoring proved too inaccurate due to discoloration of the sample from the sterile water droplet rather than from actual fungal growth. High magnification viewing provided a check on fungal reproduction (by the presence of fruiting bodies).

Table 1. Mean weight change of the polymers and resins over the 5-week exposure interval.

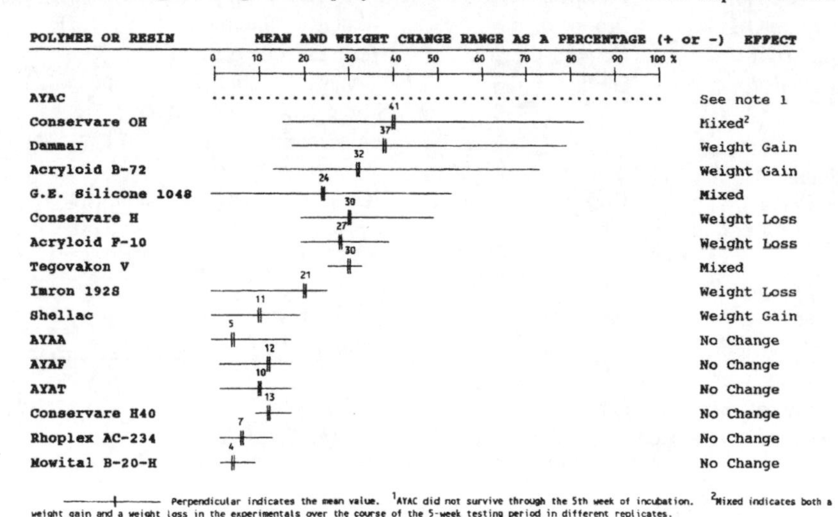

POLYMER OR RESIN	MEAN AND WEIGHT CHANGE RANGE AS A PERCENTAGE (+ or -)	EFFECT
AYAC		See note 1
Conservare OH	41	Mixed[2]
Dammar	37	Weight Gain
Acryloid B-72	32	Weight Gain
G.E. Silicone 1048	24	Mixed
Conservare H	30	Weight Loss
Acryloid F-10	27	Weight Loss
Tegovakon V	30	Mixed
Imron 1928	21	Weight Loss
Shellac	11	Weight Gain
AYAA	5	No Change
AYAF	12	No Change
AYAT	10	No Change
Conservare H40	13	No Change
Rhoplex AC-234	7	No Change
Mowital B-20-H	4	No Change

Perpendicular indicates the mean value. [1]AYAC did not survive through the 5th week of incubation. [2]Mixed indicates both a weight gain and a weight loss in the experimentals over the course of the 5-week testing period in different replicates.

2.2 Weight Loss Measurements

For the majority of the polymers and resins tested, no massive weight losses or gains were noted for control or experimental samples (see Table 1). The weight loss methods allowed for observation of large weight changes. It should be noted that some polymer and resin material was lost when scraping the material from the slide. In addition, after washing, some organisms could have adhered to the sample, resulting in higher weights associated with the experimental trials. To account for these sources of error, an arbitrary 15% weight change was considered not to be significant. Seven polymers or resins had weights that were essentially unchanged (<15%) from the control samples, 5 had a change in weight of 15-30%, 3 had a change in weight of 31-60%, and one fell apart completely. Of the materials showing a weight differential from the controls, AYAC, Conservare H, Acryloid F-10, and Imron 192S consistently had weight losses associated with the experimentals, which could signify loss caused by fungal grazing on the surface. Conversely, Dammar, Acryloid B-72, and Shellac had weight gains in the fungus-exposed test strips, which could signify water absorption, or adhesion of organism remnants on the polymer or resin during the weighing procedure.

2.3 Qualitative Fourier Transform Infrared Spectroscopy

Portions of each polymer and resin samples were analyzed by Fourier transform infrared spectroscopy (FTIR). The conclusions are summarize below:

- Silicone Resins--Most of the silicone resins showed stronger OH absorption than the controls, indicating a higher degree of hydration.
- Polyvinyl Acetates--Each of the PVACs showed different patterns of changes. However, these changes seem to be consistent within the individual polymer sets. There were apparent differences in carbonyl intensities, indicating a change in C=O bonds.
- Polyvinyl Butryl--Polymer seemed unaffected by fungal growth.
- Acrylics--Samples of Rhoplex showed variations in peak height ratios of C-H stretch bonds that correspond to chain breakage. Samples of Acryloid B-72 did not exhibit these changes. Samples of Acryloid F10 showed slight and inconsistent changes.
- Natural Resins--Dammar appeared to be unaffected by fungal exposure. While Shellac showed slight changes consistent with those expected from hydrolysis and chain scission.
- Polyimide--Imron showed some hydration of the samples.

The inconclusive nature of the results may have resulted because the chemical changes were below the sensitivity limit of FTIR, which is a 5% change and/or because the fungal growth on the samples was inhomogeneous, and portions of the sample without growth were analyzed. The results obtained, however, neither support or refute the biological effects observed in the other parts of the study. Part 2, reported below, attempted to further investigate chemical changes with FTIR.

2.4 Sporulation assessment

Assessment of the sporulation scale versus the other growth parameters used in the study showed good agreement with the RFP procedure and high magnification viewing.

As would be expected, the sporulation scale approximates the colony growth data. However, in certain consolidants such as G.E. Silicone 1048, sporulation was not evident until the fifth week of the experiment. Imron 192S, on the other hand, viewed at the fifth week of incubation showed no sporulation despite having had fruiting bodies during the first four weeks of incubation. Acryloid B-72 and Acryloid F-10 had consistently higher sporulation, which was not manifested in a greater colony growth formation.

2.5 Composite Scoring

This study developed physical and chemical methods with which to describe deterioration associated with fungal interaction on a coating surface. In the conservation field a combination of the factors of weight change, chemical change, and physical change are of importance to the conservator, over and above growth of the organism on a substrate. Thus the conservator may not care if an organism resides on the surface of a coating but would consider the effects of this residence--for example, discoloration of the object or a resultant loss of weight or a chemical change--to be of greater importance. With this in mind, a priority order was assigned to the results of the Wilcoxon's two-sample test for sporulation and colony growth, weight change, Duncan's Multiple Range Test and the FTIR results. By convention these groupings were set up for each of the above results (low, medium and high) and numbers were assigned (a 1, 2, or 3, respectively) to consolidants/resins within the groups as a weighting factor to give greater value to those effects believed to be of most importance to conservators. For example, an organism sporulating on the surface of an object would not necessarily be visible to the naked eye, or lead to colony growth; therefore it is of lower importance than a large colony growth resulting in a discoloration of the surface. The weighting scale employed gave greater emphasis to weight change, less emphasis on colony growth, and the least emphasis on the sporulation scale and the FTIR results. The results of this scoring are presented in the Table 2. In brief, the lower the composite score, the less affected the polymer or resin was by the fungi.

Table 2. Polymer and resin composite scoring.

Polymer/Resin	Rankings					
	Sporulation[a] scale	Weight[b] Change	Colony[c] Growth	Multiple[b] Range Test	FTIR[a]	Totals[e]
Rhoplex AC-234	1	3	2	3	2	11
AYAA	1	3	4	3	2	13
Tegovakon V	1	6	2	3	1	13
G.E. Silicone 1048	1	6	2	6	1	16
AYAF	2	3	4	6	2	17
Mowital B-20-H	2	3	6	6	1	18
Conservare H	2	6	4	6	1	19
Acryloid B-72	2	9	4	3	1	19
AYAT	2	3	6	6	2	19
Shellac	2	3	6	6	2	19
Conservare OH	1	9	2	6	2	20
Dammar	2	9	6	3	1	21
Imron 192S	3	6	6	6	1	22
Acryloid F-10	3	6	4	9	1	23
Conservare H40	3	3	6	9	2	23
AYAC[d]						>23

[a] Based upon a factor weight of 1.
[b] Based upon a factor weight of 3.
[c] Based upon a factor weight of 2.
[d] Final composite rank not computed since fungal treatment resulted in complete breakup of samples
[e] Lower scores indicate more resistance to fungal attack

3. DISCUSSION

3.1 Criteria for Protective Coatings

The conservation of stone monuments usually follows a standard sequence of steps (Torraca, 1975). The object being treated should be studied to determine the extent of the damage and the cause or causes of the damage or decay, with the idea of eliminating these factors if possible. The surface is then usually cleaned, with the physical or chemical removal of dirt, foreign materials or weathering crusts. Preconsolidation of the surface may be deemed advisable before cleaning if the surface is friable or in an advanced state of decay. Stone that has lost cohesion is then treated with a consolidant. Consolidation is the impregnation of the damaged areas of stone with a suitable product that reaches down into the underlying undamaged layers, and results in a strong cohesive structure. Surface protection consists of a superficial film applied to unweathered stone as a preventive measure, or applied to weathered stone after consolidation treatment. This is meant to act as a barrier to the actions of atmospheric pollutants, rainwater, or biological growths, while at the same time the surface coating should be permeable to water vapor within the stone. Surface treatments may also be effective in reducing the defacing of the surface by graffiti, by allowing spray-paints and the like to be removed more easily. A surface coating may sometimes replace consolidation when the stone surface has been eroded but the remaining stone is sound. Reconstruction, or the assembly of pieces of cleaned and consolidated stone with an adhesive, may sometimes also be needed.

3.2 Biocidal Agents

The proprietary nature of the materials tested, and manufacturer reticence, precluded a precise understanding of any potential biocidal additions in the 16 products tested, but the available information indicated that six of the polymers contained additives that may act as a biocide. One such additive is dibutyltin dilaurate, which is contained in Conservare H, Conservare OH, Conservare H40, and, perhaps, Tegovakon V. Other presumed additives that may have a biocidal effect are formaldehyde, contained in Rhoplex AC-234, and acetic acid, contained in AYAC. Proceeding from the assumption that six of the polymers and resins contained biocides, these six were examined for changes in colony growth and/or sporulation which might be associated with a perceived leaching effect of potential active biocidal agents presumed present in the polymers. Based upon this analysis, no apparent biocidal effect was noted.

Of the 10 polymers and resins that did not apparently contain any biocidal additives or contaminants, only one, G.E. Silicone 1048, showed a lag phase of sporulation over the study period. No sporulation or growth was noted over the first 4 weeks of incubation, but after week 5, an average number of fruiting bodies was noted, although with no apparent colony growth. This could indicate a short-term resistance by G.E. Silicone 1048, followed by organism growth.

3.3 Polymer and resin composite scoring

It is apparent that no polymer class behaves in a uniform manner when subjected to fungal spores; therefore, no class can be chosen based upon any one positive feature (i.e., low growth, no weight change, etc.).

For example, Table 2 shows that silicone-based polymers can be both resistant to fungal attack (e.g., Tegovakon V) and poorly resistant (e.g., Conservare H40). Due to a lack of company-supplied product data and failure of the FTIR to detect any significant bond changes, it is not possible to determine why one silicate ester is better than another; this is also generally true for the other polymer and resin classes studied. There is, however, some suggestion that during reformulations of some products, e.g., Conservare H40, up to 5% of 'inert' material may be added to 'bulk up' the product. Additives at this level would not be detected by FTIR. Further, the additives, from the manufacturers' point of view, need only prove inert to the consolidating activity of the product, not inert to microbes. It may be that a product like Tegovakon V has little or no 'inert' material added to it and, therefore, is less biodegradable than a product with 'inert' additives.

The acrylic polymer class also showed divergent behavior. Acryloid F-10 showed poor resistance to microbial growth while Acryloid B-72 showed moderate resistance. Rhoplex AC-234 ranks as the best overall in resistance to fungi.

The polyvinyls, composed of polyvinyl acetates (PVAs) and polyvinyl butryl (PVB), were interesting. As a group, the PVAs all showed moderate to good resistance to biodeterioration, AYAA, AYAF, and AYAT having composite scores of 13 to 19, with the exception of AYAC that had what we consider to be the most significant break-up of all experimental samples in that the fungus-exposed resin did not survive past week two of the experiment, while the control remained intact to the end of the 5-week test. The single sample of a PVB, Mowital B-20-H, showed better resistance to microbiological deterioration than the polyvinyl materials AYAT or AYAC. The product literature for the PVAs claims all but AYAC do not support bacterial growth. AYAC is used in the food industry as an additive, and may be expected to be biodegradable. As to the natural resins, the one used extensively in paintings conservation, Dammar, was highly degradable. Shellac, widely used in wood conservation, showed a moderate resistance to degradation.

The single polyimide, Imron 192S, fell in the poorly resistant grouping, having a composite score of 22.

Based upon the total weighted scores, those polymers or resins that showed the most overall resistance to fungal deterioration (i.e., the lowest score) were Rhoplex AC 234, Bakelite-AYAA, and Tegovakon V, with composite scores of 11, 13, and 13, respectively. G.E. Silicone 1048, Bakelite-AYAF, and Mowital B-20-H also had low weighted scores (16, 17, and 18). The polymers or resins having the poorest (or highest scores) were Bakelite-AYAC, Conservare H40, Acryloid F-10, and Imron 192S (>23, 23, 23 and 22, respectively). Other polymers or resins having a higher score were Conservare OH, and Dammar (with scores of 20 and 21, respectively). The rest of the polymers and resins (Conservare H, Acryloid B-72, Bakelite-AYAT, and Shellac) fell between the extremes, with composite scores of 19.

4. PART I, CONCLUSIONS

After considering the combined effects of organism growth, sporulation, weight change, and chemical change, it was concluded that, in regard to fungal deterioration:

- Rhoplex AC 234, Bakelite-AYAA, and Tegovakon V had high resistance.
- Conservare H40, Acryloid F-10, Imron 192S, and Dammar all showed poor resistance.
- Bakelite-AYAC was completely degraded by fungal action after 2 weeks of incubation.
- The remaining polymers and resins showed levels of resistance between those of the first two groups.

The polymers or resins that showed poor resistance to fungal deterioration, Dammar, Acryloid F-10, and Conservare H40, are extensively used in conservation today. It is therefore recommended that: (1) Conservators be made aware of the potential biodeterioration problems of these products and of any environmental controls that may reduce their susceptibility to attack; (2) usage should be restricted to 'safe' environments; and (3) these products should be replaced with more resistant products or perhaps have a biocide included within them. If a biocide is added to these products, further testing should be carried out.

5. INTRODUCTION, PART 2 FTIR OF SILICONE-BASED POLYMERS

To further elucidate the nature of the changes occurring in polymers as a result of fungal attack seven silane-based polymers currently in use in stone conservation were studied with FTIR after one and two months of fungal growth. (Analyses performed by Dr. Gretchen Shearer, The Metropolitan Museum of Art; and Dr. Peter Zanzucchi, David Sarnoff Research Labs., Princeton, NJ.).

The seven products tested were:

- Tegovakon T (Tetraethoxysilane and methyltriethoxysilane, solvents)
- Tegovakon V (Tetraethoxysilane only and solvents)
- Tetraethoxysilane
- Brethane (Methyltrimethoxysilane, solvents and catalyst)
- H40+ (more polymerized Conservare OH, plus organo-iodide biocide)
- H40 (more polymerized Conservare OH)
- DF104/B-72 (polymer silane and acrylic resin)

Samples of the above products were spread on silicon wafers and incubated for one or two months with the fungal mixture used in Part 1 (10^8

spores/ml). The samples were assayed quantitatively and qualitatively by FTIR. Two types of controls were prepared, dry and wet.

5.1 RESULTS AND DISCUSSION

The polymerization process of methyltriethoxysilane occurs in steps: an hydrolysis step with formation of a silanol, and then a condensation step when two silanols form a dimer. This reaction proceeds to form trimers, tetramers, pentamers, and so on to form larger oligomers. The more polymerization that occurs, the fewer the Si-OH molecules available for microbial attack. The presence of water is necessary for condensation to occur. This is clearly seen in the FTIR data from the wet and dry controls--the dry controls had more unreacted Si-OH than the wet controls.

Biological growth was clearly evident and extensive on all samples except Tegovakon V and Brethane. This fits in nicely with the FTIR results that showed a reduction of Si-OH groups in those products with fungal growth, i.e., Tegovakon T, H40+, H40, and DF104/B-72.

It was also felt that the weatherability of the silicones is improved with the addition of organics and the formation of copolymers (e.g., the DF104/B-72); and that since the silicone films are likely to be permeable to oxygen and water vapor they are therefore likely to be supportive of microbial activity.

ACKNOWLEDGEMENTS

This work could not have been performed with out the willing and able assistance of Mr. Edward Santoro. Ed was instrumental in devising the screening test and in statistical analysis, in addition to many hours of discussion of the topics. Mr. Mark Wypyski has made significant contributions to numerous parts of these studies for which he is sincerely thanked. Dr. Gretchen Shearer and Dr. Peter Zanzucchi spent many hours analyzing the consolidants in the second part of the study and have provided many tantalizing bits of information as to the behavior of the consolidants when attacked by fungi. Dr. A.E. Charola has been a friend and colleague who has lent a willing ear and eye to these studies. The Getty Conservation Institute kindly provided funding and intellectual input for the screening study, and The Kress Foundation kindly supported the FTIR study in part 2. And lastly, I express by warmest thanks to my editor, Vicki Koestler, who can not escape the many requests for her help in polishing my papers.

REFERENCES

Bassi, M. and D. Chiatante. 1976. The role of pigeon excrement in stone biodeterioration. Int. Biodeterioration Bulletin **12**: 73-79.

Henderson, M.E.K. and R.B. Duff. 1963. The release of metallic and silicate ions from minerals, rocks and soil by fungal activity. J. Soil Science **14**: 236-246.

Hueck van der Plas, E.H. 1968. The micro-biological deterioration of porous building materials. Int. Biodeterioration **4**: 11-28.

Koestler, R.J., Charola, A.E., Wypyski, M.T. and J.J. Lee. 1985. Microbiologically induced deterioration of dolomitic and calcitic stone as viewed by scanning electron microscopy. *In* 5th Int. Congress on Deterioration and Conservation of Stone, G. Felix, (ed.), Presses Polytechnique Romandes, Lausanne p. 617-626.

Koestler, R.J., Santoro, E.D., Preusser, F. and A. Rodarte. 1986. A note on the reaction of methyl tri-methoxy silane to mixed cultures of microorganisms. *In* Biodeterioration Research 1. C.E. O'Rear and G.C. Llewellyn (eds.) Plenum Press, New York p. 317-321.

Koestler, R.J. and E.D. Santoro. 1988. Assessment of the susceptibility to biodeterioration of selected polymers and resins. Project report to The Getty Conservation Institute, Los Angeles, CA. **xi**, p. 98

Koestler, R.J., Santoro, E.D., Druzik, J., Preusser, J., Koepp, L. and M. Derrick. 1988. Status report: Ongoing studies of the susceptibility of stone consolidants to microbiologically induced deterioration. *In* Biodeterioration 7. D.R. Houghton, R.N. Smith and H.O.W. Eggins (eds.), Elsevier Applied Science, New York p. 441-448.

Koestler, R.J., P. Brimblecombe, D. Camuffo, W. Ginell, T. Graedel, P. Leavengood, J. Petushkova, M. Steiger, C. Urzi, V. Verges-Belmin and T. Warscheid. 1994. How do environmental factors accelerate change? *In* The Science, Responsibility, and Cost of Sustaining Cultural Heritage, W.E. Krumbein, P. Brimblecombe, D.E. Cosgrove and S. Staniforth (eds.), Dahlem Workshop Report, John Wiley & Sons, New York p. 149-163.

Koestler, R.J., T. Warscheid and F.E. Nieto. 1997. Biodeterioration: Risk factors and their management. *In* Saving Our Architectural Heritage: The Conservation of Historic Stone Structures, N.S. Baer and R. Snethlage (eds.) Dahlem Workshop Report ES20, Chichester, John Wiley & Sons Ltd. New York p. 25-36.

Krumbein, W.E. 1972. Role des microorganismes dans la diagenese de la degradation des roches en place. Revue Ecologie Biologique Solai, **IX**(3). p.283.

Leznicka, S., Kuroczkin, J., Krumbein, W.E., Strzelczyk, A.B. and K. Petersen. 1991. Studies on the growth of selected fungal strains on limestones impregnated with silicone resins (Steingestiger H and Elastosil E-41). Int. Biodeterioration **28**: 91-111.

Lewis, F.J., May, E. and A.F. Bravery. 1988. Metabolic activities of bacteria isolated from building stone and their relationship to stone decay. *In* Biodeterioration 7, D.R. Houghton, R.N. Smith and H.O.W. Eggins (eds.), Elsevier Applied Science, New York p. 107-112.

May, E., Lewis, F.J., Pereira, S., Taylor, S., Seward, M.R.D. and D. Allsopp. 1993. Microbial deterioration of building stone--a review. Biodeterioration Abstracts 7: 109-123.

Nugari, M. and G.F. Priori. 1985. Resistance of acrylic polymers (Paraloid B-72, Primal AC 33) to microorganisms, First Part. *In* 5th International Congress on Deterioration and Conservation of Stone, Vol. 2, G. Felix (ed.), Presses Polytechnique Romandes, Lausanne p. 685-693.

Salvadori, O. and M.P. Nugari. 1988. The effect of microbial growth on synthetic polymers used on works of art. *In* Biodeterioration 7, D.R. Houghton, R.N. Smith and H.O.W. Eggins (eds.), Elsevier Applied Science, New York p. 424-427.

Santoro, E.D. and R.J. Koestler. 1991. A methodology for biodeterioration testing of polymers and resins. Inter. Biodeterioration. **28**: 81-92.

Smith, R.N. and L.M. Nadim. 1983. Fungal growth on inert surfaces. *In* Biodeterioration 5, T.A. Oxley and S. Barry (eds.), John Wiley & Sons, New York p. 538-547.

Strzelczyk, A.B. 1981. Stone. *In* Microbial Biodeterioration, A.H. Rose (ed.) Academic Press, New York p. 61-80.

Torraca, G. 1975. Treatment of stone in monuments: a review of principles and processes. Proc. Int. Symp. The Conservation of Stone I, Bologna p. 297-315.

Webley, D.M., Henderson, M.E.K. and J.F. Taylor. 1963. The microbiology of rocks and weathered stone. J. Soil Science. **14**: 102-112.

BIODEGRADABILITY OF PRODUCTS USED IN MONUMENTS' CONSERVATION

Piero Tiano, Lucia Biagiotti and Susanna Bracci
CNR, C.S. Opere d'Arte, Via degli Alfani 74, I-50121 Firenze, Italy

Key words: biodegradability, monuments, conservative treatments

Abstract: The conservation procedures in monuments' restoration foresee three types of interventions on stone materials: *cleaning; protecting; consolidating*. In these treatments several kinds of natural and synthetic products are applied. Most of them have a temporary function and should not remain on the monument, but some have the specific objective to protect or to consolidate the stone surface and their permanence is expected to last for a few years. For these treatments different classes of organic polymers are normally used and, in order to maintain their activity, the structure and composition of the product should remain unchanged with time. The aim of this work is to verify the rapid biodegradability of some organic products used in conservative treatments of monuments, In reality, we use the biodegradability test to verify the bioresistance of these conservative treatments, and consequently their lasting action. The readily biodegradation test has been made following the EEC method C6 "closed bottle BOD" modified for the particular field of application, using the Velp respirometer apparatus. The microorganisms used in the test have been isolated from the stones of seven monuments, located in the centre of Florence. The microoganisms have been collected using contact plates containing nutrient agar pressed on the stone surface. The strains selected are two bacteria: -ST/B3 *Micrococcus sp.*; - SC/L1 *Bacillus sp.;* and an hyphomycete: - SG/F1 *Ulocladium sp.* The restoration products used as test substances are : Paraloid B72 (acrylic resin); Silirain 50 (alkyl alcoxy silane oligomer); Akeogard CO (fluoroelastomer). The preliminary results obtained under these experimental conditions may be indicative of a low biodegradability of the products examined. Thus, it is possible that the freshly applied products possess a good bioresistance.

1. INTRODUCTION

The conservation procedures in monuments' restoration foresee three types of interventions on stone materials: *cleaning; protection; consolidation*. In these treatments several kinds of natural and synthetic products are applied (Amoroso and Fassina, 1983). Most of them have a temporary function and should not remain on the monument, but some have the specific objective to protect or consolidate the stone surface and their permanence is expected to last for a few years. For these treatments different classes of organic polymers are normally used and, in order to maintain their effectiveness, the structure and composition of the product should remain unchanged with time. At the specific site of application the protective or consolidating products can be modified principally by two mechanisms: *abiotic* or *biotic*. The former is due to physico-chemical and photochemical transformations bringing about slight modifications of the molecule, the latter by enzymes and catalysts that cause major and rapid changes.

If a compound cannot be modified enzymatically, it will not be biodegradable as the energy-yielding and biosynthetic processes cannot start (Alexander, 1994). Furthermore, the products usually applied in conservative treatments could have molecular structures to which the microorganisms have not been exposed (xenobiotic compounds). In this case, these products can be considered recalcitrant to biodegradation or subject only to biotrasformation.

With respect to the biodegradability of products applied for monuments repair and maintenance, we must take into consideration the particular ecological niche represented by a sub-aerial stone. After some years of exposure a complex biocoenosis can settle and develop on this surface, but, due to the particular environment, hardly all the organisms necessary to complete the mineralisation cycle can be present. Thus, the specific microflora may eventually carried out, on the products applied, only a partial biodegradative process.

From our point of view the absence of microorganisms able to perform a complete biodegradative pathway has a positive effect because the chemical applied can maintain its efficiency for more time. Otherwise, the biotrasformation could induce a chemical alteration of the molecular structure with a loss of some characteristic properties but without changes in the molecular complexity. This transformation could be anyway sufficient to change the product's features and, as a consequence, the effectiveness of the treatment (protective or consolidant).

The aim of this work is to verify the readily biodegradation of some organic products used in conservative treatments of monuments. This research attempts to verify the bioresistance of these conservative treatments, and, as a consequence, their lasting action.

1.1 Biodegradation

Biological processes operated by microorganisms may modify naturally occurring and many synthetic chemicals. Such modifications involve enzymes and brings about extensive changes in the structure and characteristic properties of a product. In fact, the microorganisms use these molecules as source of C, energy, N, P, S or other elements needed by the cells for biosynthetic reactions.

The complete conversion of the organic molecule to inorganic end-products is known as mineralisation or *ultimate biodegradation.* This is normally assessed by measuring the removal of Dissolved Organic Carbon (DOC) or Chemical Oxygen Demand (COD). When a chemical is biodegraded in the presence of a relatively low density of non-acclimatised microorganisms, in a relatively short period of time and in the absence of other organic compounds, it is called *readily biodegradable.* The readily biodegradation can be evaluated by determining the oxygen uptake of microorganisms when they metabolise (degrade) a product. The BOD (Biochemical Oxygen Demand) is the amount of oxygen consumed by microorganisms when metabolising a test substance, expressed as *mg of oxygen uptaken* per *mg of substance.*

1.1.1 Chemical structure

The field of study for the biodegradability is focused with the environmental pollution due to the presence of organic chemicals released into air, water and soil and that constitute a hazard for humans, animals and plants.

The microorganisms play a large role in these environments for their ability to mineralise harmful anthropogenic compounds. But we dealt with compounds that can be considered not-toxic and for their particular sector of application are only in very small amount discharged into the environment. Thus no studies on the biodegradability of such substances have been performed. In recent years, some reports have appeared suggesting the possibility that these products can be used as a substrate by the microflora dwelling on stone monuments.

Individual classes of organic compounds have different periods of persistence in soils and waters. Frequently, a slight modification of the structure of the molecule makes it more susceptible to degradation and changes the suitability of the molecule as substrate for growth by the resident community of microorganisms. Studies have been conducted on a wide variety of chemical classes providing a relationship between structure and biodegradation. The compounds chosen for these investigations have been almost exclusively pesticides, surfactants and other toxic chemicals following the needs of the particular industry involved. Some molecules are completely resistant to microbial modification and persist in nature (*recalcitrant*); usually these are undesirable since they are toxic and harmful. Moreover for some synthetic polymers (plastics and fibres) the resistance is not a toxicological issue but simply a durability concern.

1.1.2 Environmental effects

Several conditions must be satisfied for biodegradation to take place in an environment:

1. An organism must exist that has the necessary enzymes to bring about the biodegradation.
2. The organism must be present in the environment containing the chemical.
3. The chemical must be accessible to the organism having the requisite enzymes.
4. If the initial enzyme bringing about the degradation is extracellular, the site of action must be exposed for the catalyst to function.
5. If the enzyme is intracellular, the molecule must penetrate the cell membrane, or the product of an extracellular reaction must penetrate the cell for the transformation to proceed further.
6. Conditions in the environment must be conducive to proliferation of the potentially active microorganisms.

If any of these conditions cannot be met, the chemical can not be degraded. The absence of a microorganism from a particular environment, or its inability to function, means that the compound disappears very slowly or will not be destroyed at all.

During the mineralisation of an organic molecule not all of the C or other elements is necessarily converted to inorganic forms. Some may be incorporated into the cells of the active population, and some may appear as products that are typical of the conversions of naturally occurring organic materials.

Certain synthetic compounds are apparently transformed by very few species, and it is thus likely that none of the very few species with the specific enzymes may be present in a particular site. Some organic compounds are metabolised in some environment and active organisms can only be isolated from other environments.

Many biodegradations require the co-operation of more than one species. This interaction represents a sort of synergism in which two or more species carry out a transformation that one alone cannot perform.

There are specific abiotic factors that can influence the microbial transformation : temperature, pH, moisture level and salinity together with the presence of trace of organic nutrients.

The primary agents causing biological transformation in soil, water, sediments and solid surfaces are the microorganisms that inhabit these environments. The biodegradation process occurs at a faster rate under aerobic conditions when appropriate environmental conditions, nutrient and microorganisms are present.

The metabolic pathways and environmental conditions required to achieve biodegradation of these products are unique to the particular compound (Karsa and Porte, 1995).

1.1.3 Acclimation

Prior to degradation of many organic compounds a period occurs in which no destruction of the chemical is evident. This preliminary period is the length of time occurring between the addition of a chemical into an environment and evidence of its detectable disappearance. The length of the acclimation period depends on the chemical's concentration, the environment, the temperature, the aeration status and other, often undefined, factors. As a consequence it is very difficult to predict the duration of this phase. The acclimation phase may be longer in environments in which the concentration of N, P, and other inorganic nutrients is low. Furthermore, this could reflect the period of time required for a mutation or gene transfer to occur, or for the biosynthesis of inducible enzymes in appreciable amounts. In the latter case, a *threshold* of the substrate exists for the induction of certain enzymes.

The acclimation could be simply the result of the time required for a small population to become sufficiently large to give a detectable loss of the organic substrate.

Presently, it is not possible to predict which biodegradable chemical in what environments will persist because of the threshold and which will be destroyed because of the ability of the responsible populations to function at still lower levels of the substrate.

This acclimation phase is normally considered detrimental since the exposure period can be harmful for human health but, in our case, it is beneficial as the product remains unaltered and can maintain its properties for longer periods of time.

If an organic compound is persisting it is evidence that:

- microorganisms are not acting;
- they are acting very slowly;
- no microorganisms exist with the capacity to modify the molecules.

1.1.4 Inoculum

Previous experience has shown that the larger the size of the inoculum the shorter the time required to degrade a chemical. The nature and the quantity of the inoculum play an important role in biodegradability assessments; the inoculum is probably the biggest single factor in the success of the test. Pure cultures or mixed populations can give very different results. We deal with a very specific sector and the usual sources of inoculum such as river water, sewage effluent, activated sludge and soil (Walter and Crawford, 1997) are not suitable for our purpose. In fact, no useful information on the biodegradability (bioresistance) of conservation products can be achieved using inoculum coming from unrealistic exposure of the product to be tested. Consequently, we use as inoculum pure cultures of microorganisms dwelling on the surface of stone monuments.

2. EXPERIMENTAL

The readily biodegradation test has been made following the EEC method C6 "closed bottle BOD" modified for the particular field of application.

2.1 Monuments

The microorganisms used in the test have been collected, using contact plates (containing nutrient agar), from seven monumental stone surfaces, located in the centre of Florence. The lithotype of all the samples was Pietraforte, a typical sandstone quarried near Florence and used mostly for the construction of monuments and religious buildings. From the numerous colonies developing on the contact plates after 5 days (Table 1) bacteria and fungal strains with different macroscopic aspect have been selected and sub-cultured.

Table 1 Microorganisms isolated from Florentine monuments

Code	Name	Sampling area	N° of colonies	Strains isolated
LL	Loggia dei Lanzi	Basement, on the left of the entrance	103 bacterial 39 fungal	F4 – L3 – B7
SC	S. Carlo Church	Right side of the door	230 bacterial 35 fungal	F3 – L2
ST	S. Trinita Church	On the flat column between left and central doors	28 bacterial 4 fungal	F2 – B3 – B4
PA	Antinori Palace	Left corner, 3rd row, 4th slab	37 bacterial 2 fungal	B2
SG	S. Gaetano Churh	Left column, left door	6 bacterial 8 fungal	F1
PS	Strozzi Palace	Under 3rd ring, right side of entrance	62 bacterial 25 fungal	F5 – B5 – B6
SM	S. Maria Maggiore Church	Right column of the door	85 bacterial 33 fungal	B1 – L1

2.2 Test organisms

The following bacterial and fungal strains isolated from contact plates were used for the BOD_5 tests:

- **ST/B3** – Light pink bacterial colony, formed by small $Gram^+$ cocci, genus *Micrococcus*
- **SC/L1** – Cream bacterial colony, formed by $Gram^+$ bacilli, genus *Bacillus*
- **SG/F1** – Hyphomycete, dark conidiophores, ovate porospores, genus *Ulocladium*

A portion of each strain's colony from nutrient agar plates was suspended in 5 ml of physiological sterile solution (PSS). An aliquot of this suspension was then added to 14 ml of PSS to give a final concentration of about 10^6 cell/ml (by optical density measurements). The inoculum for each bottle was adjusted to a final volume of 150 ml with PSS.

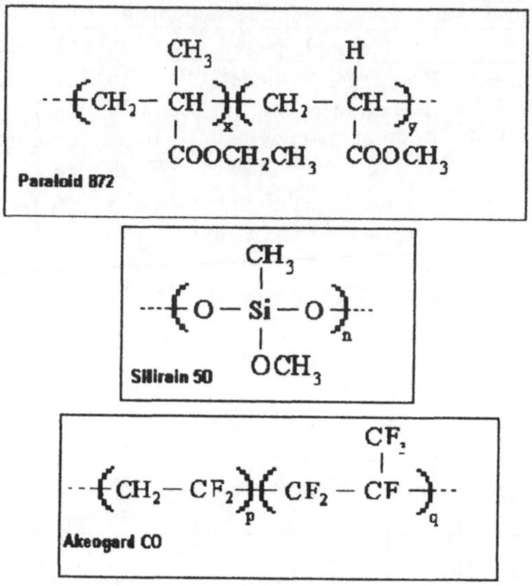

Paraloid B72

Silirain 50

Akeogard CO

Figure 1. Chemical structure of the test products.

2.3 Test Products

We have used as test substances some of the products most commonly utilised for the protection and the consolidation of monuments. Their chemical structure is reported in *Figure* 1.

a) *Paraloid B72* (acrylic resin); protective
b) *Silirain 50* (alkyl alcoxy silane oligomer); consolidant
c) *Akeogard CO* (fluoroelastomer); protective, superficial consolidant
d) *Positive control* (22.5 mg/l of dextrose and 22.5 mg/l of glutamic acid) (BOD_5 esteemed 220 ± 18 mg O_2/ for 15 mg/l of both at 20°C).

As the original products are liquid formulations, an aliquot of each of them was poured into a Teflon Petri dish (10 cm) and kept at room temperature until complete evaporation of the solvent and formation of a thin solid film. This film was finely ground with mortar and pestle; an equal portion of 80 mg of each product was used for the various tests.

2.4 Test Apparatus

The BOD_5 determination has been made using the Velp Scientific respirometer equipped with 6 bottles (150 ml). The test samples have been incubated for 5 days at 20°C with continuous agitation.

2.5 BOD_5 Tests

[I]

1. Product A + ST/B3
2. Product A + SC/L1
3. Product A + SG/F1
4. Product B + SC/L1
5. Product B + ST/B3
6. Positive control + ST/B3

[II]

Product B + SG/F1
Product C + ST/B3
Product C + SC/L1
Product C + SG/F1
Product A + (ST/B3 – SC/L1 – SG/F1)
Positive control + SC/L1

[III]

1. Product B + (ST/B3 – SC/L1 – SG/F1)
2. Product C + (ST/B3 – SC/L1 – SG/F1)
3. Product A
4. Product B
5. Product C
6. Positive control + SG/F1

The different combinations (product/inoculum) have been tested 3 times (the standard deviation is below 10%).

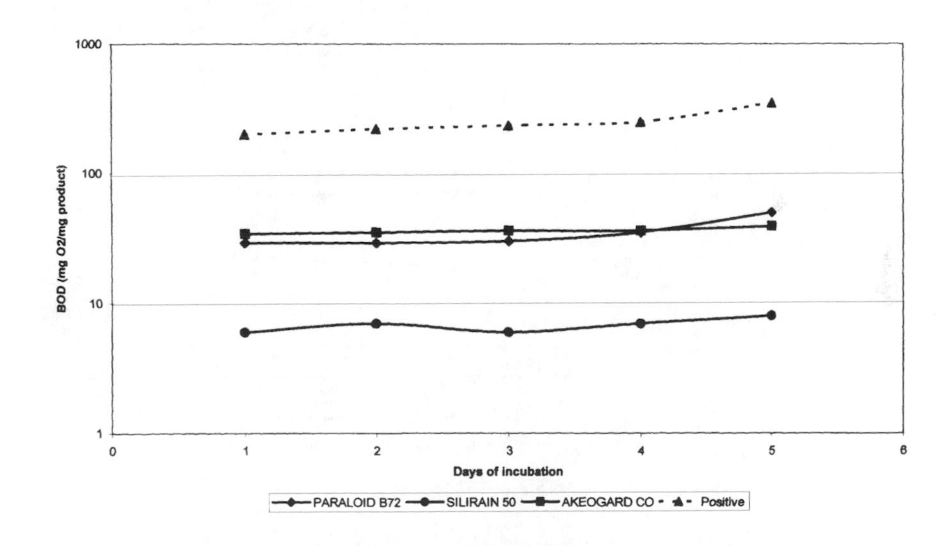

Figure 2. Biodegradability test made with the *Micrococcus* sp. (strain ST/B3)

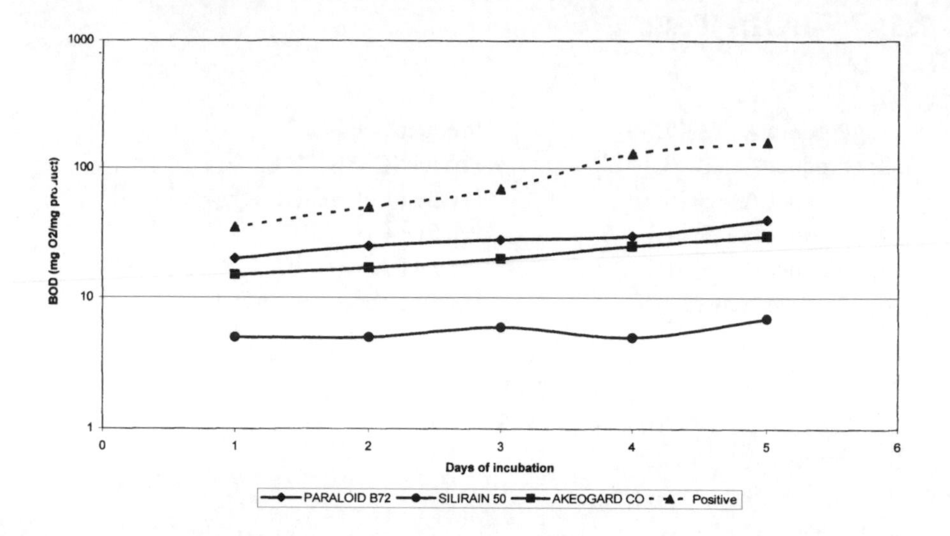

Figure 3. Biodegradability test made with the *Bacillus* sp. (strain SC/L1)

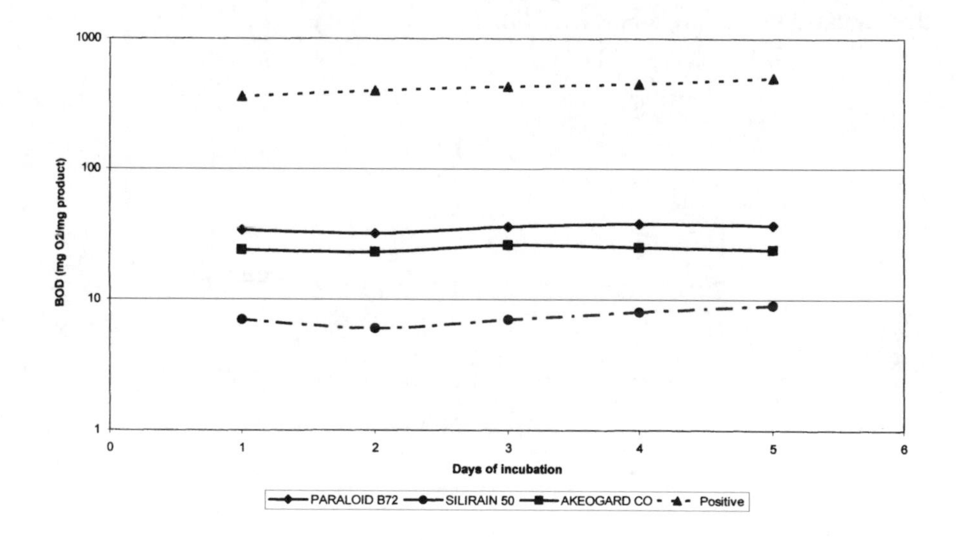

Figure 4. Biodegradability test made with the *Ulocladium* sp. (strain SG/F1)

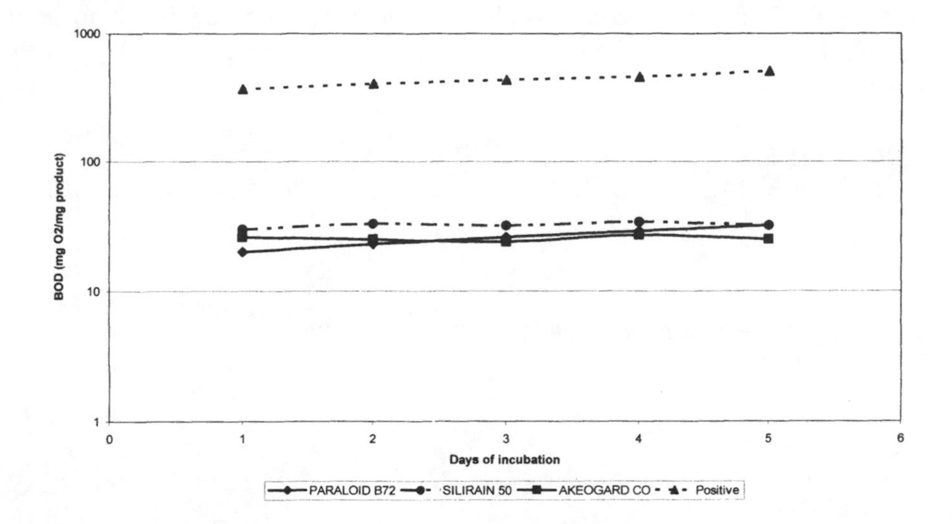

Figure 5. Biodegradability test made with the mixed population (strains ST/B3-SC/L1-SG/F1)

2.6 Results

The results of the BOD_5 test are reported in Table 2 as mg of O_2 consumed for mg of product. These are grouped and plotted (*Figure*s 2-5) by the test organism used in the various tests. As it is possible to note, none of the microorganisms used (or their mixture) has shown significant consumption of O_2 thus indicating that biodegradation of the products tested has probably not occurred.

3. DISCUSSION

The data obtained with this first biodegradability test, made on some products used for monuments conservation, seem to indicate a marked resistance of these chemicals to microbial attack. The results could be attributed either to a recalcitrant nature of these molecules or to the absence of biodegradative microorganisms in the samples collected from stone surfaces. Naturally, the particular environment and the molecules interested in the process are quite unusual for this kind of study and further experiments are necessary to evaluate the effective biodegradability of such types of compounds. As previously stated, the type of the "inoculum" can greatly influence the results of the test; as a consequence, the use of different strains, alone or mixed, could be more successful. Moreover, in these experiments we tested only freshly applied chemicals, whereas it could be advisable to let a certain period of time elapse after the application, before checking their biodegradability.

Table 2 BOD_5 results

Product/Strain	1st day	2nd day	3rd day	4th day	5th day
A – PARALOID B72					
ST/B3	29	29	30	35	50
SC/L1	20	25	28	30	40
SG/F1	34	32	36	38	37
ST/B3 – SC/L1 – SG/F1	20	23	26	29	32
B – SILIRAIN 50					
ST/B3	6	7	6	7	8
SC/L1	5	5	6	5	7
SG/F1	7	6	7	8	9
ST/B3 – SC/L1 – SG/F1	30	33	32	34	32
C – AKEOGARD CO					
ST/B3	34	35	36	36	39
SC/L1	15	17	20	25	30
SG/F1	24	23	26	25	24
ST/B3 – SC/L1 – SG/F1	26	25	24	27	25
D – Positive Control					
ST/B3	200	220	235	250	350
SC/L1	35	50	69	130	160
SG/F1	360	400	430	450	500
ST/B3 – SC/L1 – SG/F1	370	405	435	460	510

In fact, once applied to the monument, physico-chemical or photochemical reactions can transform the structure of the chemicals. This transformation, together with the enrichment in growth factors of treated stone substrata, could be necessary to start the biodegradative process. In such a case the microorganisms involved maybe be the same that seemed to have failed in the tests made on fresh products. But, in this case, we must consider that the possible transformation of the molecules could be sufficient to alter their characteristics (e.g., loss of the water-repellence or of the consolidating effect) and, as a consequence, their performance. Thus, after an acclimation period, the treatment could not be more efficient on stone surfaces and the assessment of the molecules' biodegradability not more needed for evaluating their bioresistance.

REFERENCES

Alexander M. 1994. Biodegradation and Bioremediation. Academic Press, New York p 9-10

Amoroso G.G. and V. Fassina. 1983. Stone Decay and Conservation. Material Science Monographs 11, Elsevier, The Netherlands.

Karsa, D.R. and M.R. Porte (eds.) 1995. Biodegradability of surfactants, Blackie Academic & Professional, London p 67-73

Walter, M. V. and R.L. Crawford. 1997. Overview: Biotrasformation and Biodegradation. *In* Manual of Environmental Microbiology, H. Knudens and W. Stetrenbach (eds.), ASM Press; Washington, D.C. p 707-708

PART 3

CONTROL AND UTILIZATION OF MICROORGANISMS

INTEGRATED CONCEPTS FOR THE PROTECTION OF CULTURAL ARTIFACTS AGAINST BIODETERIORATION

Thomas Warscheid
IWT- Institute for Material Science - Microbiology -, Paul-Feller-Str. 1, D-28199 Bremen, Germany

Key words: biodeterioration, interdisciplinarity, protection, cultural artefacts.

Abstract: The importance of microbial impacts in the alteration and deterioration of cultural artefacts made of mineral, metallic or organic materials has been widely acknowledged in the course of many recent investigations. While in the past biodeterioration problems on cultural artefacts were often handled without a detailed analysis and, as a consequence, simply controlled by biocidal treatments, a much deeper interdisciplinary understanding of the environmental factors and material properties regulating the biogenic damage factor would allow more specific and adequate interventions. The upcoming possibilities and potential realisation of integrated concepts in the protection of cultural artefacts against biodeterioration will be basically explained with the analytical and evaluating approach and will be exemplified by the presentation of case studies dealing with the conservation of historical objects made of natural stones, wall-paintings, historical glass, archives (e.g. paper, leather, parchment) and organic coatings.

1. BIODETERIORATION MECHANISMS

Whether as single or catalytic enhancing factor, microorganisms, like algae, cyanobacteria, lichens, bacteria and fungi, influence, due to their contamination, growth and metabolic activity, the complex interaction between various types of materials and the surrounding physical as well as chemical damage functions (Koestler et al., 1992).

In the course of biofouling (e.g. presence of colloidal microbial biofilms on or inside the materials), the microflora leads, besides an aesthetical impairing due to biogenic pigments (e.g. green chlorophyll, brownish melanin, red carotenoids), to the alteration of physico-chemical characteristics of the materials with regard to their (i) mechanical properties, (ii) surficial absorbency/hydrophobicity, (iii) diffusivity and (iv) thermal-hygric behaviour (Warscheid, 1996a).

Subsequently, the microbial consortia may cause a biocorrosive attack (e.g. microbial induced or influenced corrosion of materials) leading to the alteration of the structure and stability of materials by the (i) phototrophic enrichment of organic biomass, (ii) the selective cellular enrichment and redox processes of cations and anions (e.g. iron, manganese), (iii) the excretion of corrosive metabolic products (e.g. organic and inorganic acids) as far as the (iv) enzymatic mineralization of organic materials (Warscheid and Krumbein, 1996).

Over and above that germs, spores, dead cells and microbial toxins (e.g. endotoxins, mycotoxins) possess an allergenic pathological capability affecting restorers and conservators as well as visitors and users of cultural artifacts, especially in libraries and archives (Gallo, 1993; Hödl, 1994a).

2. MICROBIOLOGICAL MATERIAL ANALYSIS

The attention to biodeterioration problems of people in charge of restoration and conservation of cultural heritage has revealed a growing demand for an entire evaluation of the importance of microbial impacts interacting with material-immanent properties as well as natural and anthropogenic influences during the deterioration process (Koestler et al., 1997). The consistent analytical approach comprises in the order:

- object anamnesis (e.g. damage description, object history, climatic/ environmental conditions, material properties, former protective treatments),
- non-destructive observations (e.g. videomicroscopy, remission spectroscopy, respiration/photosynthesis measurement, assessment of ATP-content),
- microscopical studies (e.g. biofilm staining procedures (PAS/FDA), light and fluorescence microscopy, SEM, CLSM),
- biochemical measurements (e.g. quantification of proteins / phospholipids as biomass, analysis of pigments) and finally
- microbiological investigations (e.g. enumeration of air-borne and material-immanent microorganisms, characterization and taxonomical classification of the microflora, simulations studies, toxicological studies), according to

the analytical strategies proposed by May and Lewis (1988) as well as Becker et al. (1994).

Over and above that, the biodeteriorating effects need to be proven by a quantification of complementary changes in the material properties (e.g. discolouration, loss of weight, weakened stability, increased roughness, altered structure/porosity, increased absorbency/hydrophobicity). In this course changes in the physico-chemical behaviour of the material to the environment should be addressed, such as its thermal-hygric behaviour due to the darkening of the material surface by biogenic pigments, its tendency for an increased deposition of pollutants due to the presence of a sticky biofilm and its alteration in the moisture transport due to the effects of pore filling biofilms (Warscheid et al., 1991; Warscheid, 1996b). In specific cases, the potential hazardous risks of microbial metabolites to human health (e.g. allergenic spores, toxins, pathogenic microorganisms) should be considered, analyzed and evaluated complementary (Hödl, 1994a; Averdieck et al., 1997; Flannigan, 1997).

3. ENVIRONMENTAL CONDITIONS FOR BIODETERIORATION PROCESSES

During the anamnesis of biodeterioration processes of cultural artifacts it is important to register the environmental conditions which are favourable for a microbial infection, contamination and biodeterioration process in particular, in order to consider and establish already here the basic parameters for effective countermeasure strategies.

3.1 Biofilm - a stabilising microniche-

It is important to stress the fact that a material-specifc microflora is embedded preferably in a colloidal slime layer, called biofilm. Due to the presence of colloidal polymeric substances, the biofilm protects the microorganisms by balancing changes in humidity and temperature as well as osmotic- as well as pH-relevant influences. Based on its considerable ion-exchange capacity, the biofilm even impairs the penetration of biocides, detergents or antibiotics preventing the control of the microbial contamination and biodeterioration processes in the long-term.

Over and above that, the arrangement of microbial consortia in a biofilm matrix leads to the stimulation of their metabolic activity by (i) the extension of the colonization area, (ii) the deposition and enrichment of nutrients on the adhesive surface, (iii) the promotion of a microbial metabolic network ("cross-feeding") and (iv) the support of intracellular communication by the exchange of genetic information. Therefore, in contrast to medical microbiology, the "pathogenic" effects of microorganisms on materials only rarely are due to the activity of one species, but are more often caused by complex microbial consortia characterised by a high adaptability and flexibility during the biodeterioration process (Flemming and Schaule, 1994; Warscheid, 1996a).

3.2 Exogenic Parameters

The microbial contamination on and in materials is basically determined by the availability of water provided by rainwater, rising dampness and condensational moisture, depending on the sorption isotherms of the respective material. Fungal growth will be possible within a water activity (e.g. ratio of the vapor pressure of the air in equilibrium with a substance or solution divided by the vapor pressure at the same temperature of pure water) of $a_w > 0,6$ and a time of wetness TOW > 0,5 (e.g. more than 12 h during a day); optimal conditions for their growth will be within an $a_w \sim 0,75$ (Adan, 1994). Other microorganisms, such as algae or bacteria, probably need higher moisture supply ($a_w > 0,9$), but, in the widespread presence of moisture-conserving biofilms, these microorganisms may survive in infected materials even under more unfavourable moisture conditions (Flemming and Schaule, 1994).

In the long-term, the material structure (e.g. surface roughness, absorbency/hydrophobicity, porosity and inner surface) determines the adhesion, colonization and spreading of the microorganisms on and within the material (Warscheid et al., 1993). In addition, it may further support the microbial by providing inorganic and organic nutrients. Further decomposable nutrient sources may be offered by the exposure to light, leading to the enrichment of photosynthetic biomass, as well as by the deposition of natural and anthropogenic aerosols (e.g. ammonia, nitrate or combustion hydrocarbons; Warscheid et al., 1991; Warscheid et al., 1993; Saiz-Jimenez, 1995; Mitchell and Ji-Dong., 1999). To evaluate the nutritive conditions for a particular microbial consortium, it is important to consider that microorganisms settling on material surfaces are able to survive or even grow under oligotrophic conditions (i.e. low concentrations of nutrients) (May et al., 1993). The contamination process will even be extended when the material possesses buffering capacities for biogenic metabolic compounds with acidic properties, since the optimum of pH for most of the microorganisms present on cultural artifacts is around neutrality.

The optimal temperature for most of the microorganisms involved in the biodeterioration of cultural artifacts ranges between 16 and 35°C. The oxygen supply will not exclude microbial activity, but will determine the type of metabolic pathway to be followed (oxidative or fermentative). Finally, the possible ways of contamination (e.g. air-borne, infected materials) have to be analyzed and considered as potential cause of microbial infections and biodeterioration processes on historical objects.

4. PROTECTIVE TREATMENTS IN GENERAL

The control of biodeterioration processes on materials will be basically achieved by effective measures to limit and restrict the above mentioned growth conditions for the respective microflora. Therefore it is preferable "good house-keeping" and climate control as well as the application of appropriate, especially microbial resistent, protectives before applying ecotoxicological questionable and dangerous biocides (Warscheid and Kuroczkin, 1997).

4.1 "Goodhouse-Keeping" and Climate Control

The protection of cultural artifacts from biodeterioration will be primarily achieved by the reduction of moisture in and around the endangered object in by drying, drainage, ventilation, climatized storage. Due to the moisture-conserving effects of microbial biofilms, artifacts already contaminated should be kept probably more safe at humidity levels lower than 55 r.H. %, whereas non-contaminated objects will tollerate up to 65 r.H. %, depending on the type of material and its sorption isotherm.

The growth-controlling effect of desiccation will be increased by a subsequent soft cleaning (e.g. vacuum cleaner, brushes) and, if necessary, careful disinfection (e.g. medical alcohol, if the case combined with conserving agents) of the material surface removing moisture-absorbing dust, attached particles, crusts or biogenic slimes. In this context, it has to be emphasized that, during cleaning, the input of additional moisture should be limited as much as possible (e.g. by wringed-out cloths, fine-part dry cleaning), otherwise the cleaning effect will remain only temporarily until the remaining microflora recovered in the moist condition (Warscheid et al., 1988).

The long-term effect of these measures will be extended as long as the further contamination of the objects by aerosols, particles and microorgamisms can be avoided. This will require the formulation and realization of technical guidelines for a "good house-keeping" (i.e. regular cleaning and repairing, appropriate heating and ventilation). Modifications in the illumination, temperature conditions, pH-range or redoxpotential of the affected materials and their environment will have only slight inhibiting effects against the mostly high adaptable microflora, not mentioning the problems of their practical realization.

Nevertheless, already this practical proceedures will provide sufficient relief in most cases of biodeterioration of cultural artifacts, even when microbially sensitive materials are present, and the individual steps can be easily integrated into the general conservation work scheme (Warscheid and Kuroczkin, 1997).

4.2 Selection of Protectives

The application of protective agents, such as coatings, consolidants, water repellants, fillers as well as fixatives and organic binders, should primarily be carried out with respect to the prevailing physico-chemical conditions of the object material and its refering damage situation (Wendler, 1997). If microbial contamination and bio-deterioration processes are clearly proven, the selection of appropriate protectives should consider their microbial resistance in order to avoid the initiation, reoccurence or even acceleration of microbial damage of the cultural objects under study (Koestler, 1999; Tiano et al., 1999).

The microbial resistance of materials should be preferably tested in the laboratory with material-specific microbial consortia as well as in situ on the cultural object in question (von Plehwe-Leisen et al., 1996; Warscheid and Kuroczkin, 1997).

4.3 Biocidal Treatments

In the more serious cases of biodeterioration, where the possible improvement of given materialspecific, expositional and environmental conditions are inevitably limited and cannot be changed, further countermeasures will be required. In order to increase the durability of restoration and conservation treatments on cultural artifacts heavily affected by biodeterioration processes, the use of biocides as additives might be unavoidable (Kumar and Kumar, 1999).

Antimicrobial active substances can be commonly distinguished between alcohols, aldehydes, organic acids, carbonacidesters, phenols and their derivates, halogenated compounds, metals and metal organic substances, oxidative compounds, enzymes, surface-active compounds and various synthetic organic products.

Commercial biocide products used in the classical conservation with a considerable rapid effect comprise mainly:

- quaternary ammonium compounds,
- chlorine or halogenated compounds,

- metall-organic compounds
- aromatic compounds (e.g. phenols, formaldehyde, CMK) or
- isothiazol-derivates.

Alternative active substances, which because of their low toxicity are more easy to handle in the conservation practice, are:

- metallic salts (e.g. copper, zinc; Richardson, 1988),
- acetic or salicylic acid (Kumar and Kumar, 1999)
- borax ("polybor") (Richardson, 1988)
- PHB-esters (Lesznicka, 1992; Hödl, 1994b)
- ethereal oils (e.g. thyme oil, rose oil, neem-oil), or
- pyrethrum.

Of course, the application of these alternatives requires more time before the killing effect is obtained. Over and above that, it takes for granted that the microbial infection of the treated object will be monitored and a sufficient after care will be established; ideally this should be the case for every cultural artifact after restoration and conservation !

The formulation of synergistic working biocidal agents could probably result in an increase of their effectiveness whilst reducing the chemical load to the contaminated material as well the toxicological risk for the restorers and conservators applying the treatment. Here, a major gap in the knowledge is obvious and needs to be filled by further ecotoxicological research.

Since the impact of ionizing radiation or UV-irradiation will be probably limited to an antiseptic treatment of surfaces, a complete sterilization of highly infected cultural artifacts will be obtained by the application of ethylenoxide in computer controlled chambers.

It has to be emphazised that the theoretical effectiveness of antimicrobial substances might be strongly reduced in practice due to the physiological and ecological flexibility of the microbial consortia embedded in the colloidal biofilm on the respective materials in situ as mentioned above (Tiano et al., 1995). Especially the cometabolic organization of the respective microflora makes organic synthetic structured biocides in the long-term highly susceptible as nutritive substrate, reestablishing thus the preceeding biodeterioration processes. As a consequence and with regard to the design and use of microbicides, proof of their effectiveness against a broad spectrum of material-specific microorganisms has to be given in order to avoid later selectivity, adaptation and resistance of the prevailing microorganisms. Over and above that, any possible detrimental side-effects of the treated material by colour changes, corrosion or internal crystallization has to be avoided and preliminary tested in laboratory studies (Wakefield and Jones,1996; Nugari, 1999). Finally, ecotoxicological considerations demand a careful use of biocidal additives in order to limit possible health risks for the applying conservator as well as the later visitors of the cultural artifacts (Warscheid and Kurozckin, 1997).

5. INTEGRATED CONCEPTS IN EXEMPLARY CASE STUDIES

The assessment of the biodeterioration damage of cultural artifacts, whether archived in-doors or openly exposed, meaning the clear prove and differential diagnosis of microbial impacts within the actual deterioration process, will necessarily demand the development of integrated concepts with respect to a long-term prevention of the respective objects. The benefit of an interdisciplinary and complementary cooperation of conservators and microbiologists in the evaluation and handling of the effects of biodeterioration on cultural artifacts will be demonstrated in the following case studies based on recent research activities of our institute with the conservation practice.

5.1 Temple of Angkor Vat, Cambodia (stone)

The complex of the temple of Angkor is located near the town of Siam Reap close to the lake "Tonle Sap" in the center of Cambodia. The region lies in a tropical climate with intensive dry and rainy seasons. The buildings were built between the years 802 and 1295. The assemblage of temples represents the largest religious monument of the world; more than 100 temples are disseminated over an area of 230 square kilometers.

The studies for conservation of "Apsara"-reliefs, described here, were realized in the largest temple of the Angkor complex "Angkor Vat" during a research project of the restoration and conservation division of the Polytechnic University of Cologne (Germany). Due to a widespread corrosion and scaling of the sandstone, the "Apsara"-reliefs are in danger of becoming unretrievably lost. In order to develop an effective conservation strategy, it was necessary to analyse the causes of the stone deterioration and to evaluate the influence of the microbial impact.

The macroscopical anamnesis of the biodeterioration of the Angkor Vat and its ornamental "Apsara"-reliefs showed that, in the original state, the stone material was almost completely covered by algae and lichens; some distinct red and black areas could even be seen. During the biocidal cleaning by an Indian conservation project in the early 90's, the surficial biogenic contamination was removed, leading to a deep blackening of the treated stone surfaces in the following years. Besides this, some parts of the temple showed an intensive iron oxidation within the uppermost layers of the stone.

The microbiological assessment of the biodeterioration status comprised the microscopical analysis of the microbial contamination in the stone profile of different damaged forms after PAS staining for microbial biofilms, the cultural enrichment of phototrophic and heterotrophic microorganisms and the quantification of their metabolic activity by respiration and photosynthesis measurements as well as ATP analysis. In short, the data revealed a very slight biocorrosive activity by lichens and iron-oxidizing bacteria. The blackening of the treated stone surfaces could be attributed to melanin-producing fungi, whose growth was probably encouraged by the removal of the lichens years before. Considering the thermal effect of blackened surfaces, especially under tropical climates, adding hygric stress to the affected stone surface, the up-coming contamination by black fungi has to be considered up to this point as a major threat for the stone material than the former biopatina of algae and lichens.

For these reasons, an effective conservation strategy for the "Apsara"-reliefs should include biocidal treatments only, if a control of the entire microbial community present at the Angkor Vat will be established. Synergistic treatments, including the oxidative destabilisation of the microbial biofilms by hydrogen peroxide, soft mechanical cleaning of the stone surface and the subsequent application of metallic salts, based on copper and zinc, with depot function offered the most positive effects so far. Over and above that, the widespread contamination by fungi requires necessarily the application of microbially resistant consolidants in order to extend the durability of the proposed conservation treatment.

5.2 "Allerheiligen" Chapel in Wienhausen, Germany (wall-painting)

The monastery of Wienhausen is located in the western part of Lower-Saxony in Germany. The studies for conservation concentrated on the frescoes in the "Allerheiligen" Chapel integrated in the gallery of the monastery. The study was made in cooperation with the Lower-saxonian Institute for Preservation of Historic Buildings and Monuments in Hannover (Schwarz, 1996).

During the macroscopical anamnesis of the frescoes, dated from the 13th century, a grey patina and intensive sanding on the painted layers could be observed; furthermore a surficial fungal contamination was visible. Red stains could be seen in parts of the wall painting probably burdened with heavy salt incrustations (e.g. sulfate and (hygroscopic) nitrate). The humidity levels in the chapel changed dramatically between 45 and 85 % r.H., favouring salt crystallisation or microbial growth.

The status of the biodeterioration was assessed by means of non-destructive analytical tools such as fluorescence- and videomicroscopy to describe the extent of the microbial infection, Rodac impression plates for isolating the surficial microflora and ATP-analysis in order to quantify the microbial activity *in situ*. Furthermore, a testfield for the application of fixatives, including Syton X 30 (colloidal silica in aq. dest., 3%), Acronal 500 D (acrylate dispersion in aq. dest., 5%), Klucel EF (hydroxypropylcellulose in EtOH, 2,5%) and Klucel GF (hydroxypropylcellulose in EtOH, 2,5%) was prepared by the restorer in charge of this object.

The microbiological analysis proved the presence of biocorrosive bacteria and some fungi on the sandy painted layer of the fresco, finding nutritive sources in a former casein fixation. Furthermore the data gave indications that the red stains could be attributed to the growth of halotolerant bacteria. Long-term observations of the fixative testfield showed that the application of even highly bioreceptive cellulose fixatives (Klucel) in medical alcohol resulted in a considerable relieve of the microbial infection and no reoccurrence of active fungi could be observed nearly one year after the desinfectant fixation. In contrast, on the application of the supposed microbially more resistent colloidal silica fixatives (Syton X 30), the microbial contamination seemed to be unaffected, probably because the application was made in pure water, stabilising the microbial infection without controlling (disinfectant) effect.

Considering that the application of Klucel implied a reversibility of the fixative treatment, the application of this substance in the specific case was chosen by the responsible restorer. In order to avoid a potential microbial attack of the potentially degradable fixative in the long-term, a regular monitoring of the object was recommended, besides a subsequent climate control (< 60 % r.H.).

5.3 "Cosmas and Damian" Church in Goslar, Germany (glass paintings)

During a restoration project with the glass conservation center in the cathedral of Cologne, romanic glass paintings, called "Sechspass" and "Angel" from the "Cosmas and Damian" church of Goslar, were analysed for their black-brownish layers and surficial incrustations on selected glass fragments. Under the incrusted surfaces, but also partly on the glass surface itself, intensive fungal growth could be seen even macroscopically. The fungal contamination was probably caused by a moist and non aereated storage of the glass paintings, wrapped in newspapers and kept in a wooden frame. Under these conditions, fungi grew utilizing an oil or wax varnish applied in the past. Given the obvious microbial contamination, the immediate removal of the newspaper and a ventilated storage of the glass paintings was advised as a first step.

The microbiological analysis of the glass paintings comprised again non-destructive methods, such as fluorescence-videomicroscopy in order to describe the distribution of the microbial contamination, Rodac impression plates to isolate the relevant biodeteriogens and the subsequent analysis of their biocorrosive potential, characterised by the excretion of organic acids or the oxidation of metals (e.g. iron, manganese). The analytical data pointed to the presence of microbial biofilms as catalysers of the subsequent inscrustation process mainly caused by the increased deposition of acidic pollutants from the surrounding environment. The browning of selected glass fragments could hardly be attributed to the microbial activity.

As a consequence, the conservation work was confined to a soft mechanical cleaning and removal of the incrustation combined with a disinfection with medical alcohol. The application of protective glass coverings at the building site and a regular monitoring of the object should secure the long term effect of the suggested measures.

5.4 Public Archives in Trier, Germany (paper, leather, parchment)

In the case study presented here, about ten percent of a stock of 385,000 books in a public archive were found to be highly infected by fungi, chemoorganotrophic bacteria and actinomycetes and visibly damaged by discoloration ("foxing") as well as by enzymatic and biocorrosive disintegration. Over and above that allergic symptoms were reported, probably due to germs, spores, dead cells and microbial toxins (e.g. endotoxins, mycotoxins). The microbial contamination was mainly caused by an insufficient maintenance of the book, as well as micro- and macroclimatic problems within and around the archives. The biodeterioration problems were probably intensified and, therefore, more obvious for the inclusion of a highly contaminated book collection coming from private hands.

The non-destructive microbiological analysis of the contaminated material comprised videomicroscopy, respiration measurements, ATP-analysis, Rodac impression plates as well as sampling of air-borne microorganisms and toxins. The results of these investigations revealed a rather high contamination by air-borne microorganisms in several rooms of the archives. The microbial contamination of books was mostly restricted to outer parts, such as cover, spine and paper line. However, the composition of the isolated microflora was mostly inconspicuous, and correlations with the observed deterioration phenomena could be found. Isolated fungi showed a high cellulase activity, while the microbial attack on collagen fibers of parchments could be attributed to the presence of streptomycetes. Some of the books analysed showed visually distinct contamination areas (e.g. "foxing spots"), which could not be clearly differentiated by cultural techniques, but associated to microorganisms by ATP-analysis.

For the preventive and protective treatment of the contaminated books and archives rooms, first of all regulations concerning the handling of infected archive materials and the behaviour of employees in contaminated archives were formulated. The empty archive rooms were then cleaned throughly using a special vacuum cleaner (type K1) and the shelves were wiped out with a household cleaner. The climatic conditions in the archives were improved by changes in the building construction (e.g. roof, windows, plaster) and by the installation of an effective air-conditioning system. After drying the contaminated books, the subsequent precleaning of visibly infected areas in and above the archive materials was realized by a vacuum cleaner or the use of soft brushes under clean benches. Finally, in selected cases the book surfaces were cleaned additionally with a cloth slightly soaked with a solution of PHB-Esters (1,5 % in iso-propanol) or borax (5-10 % in tap water), the later being mainly used on objects made of parchment. In order to avoid the irretrievable loss of heavily infected objects, certain books were proposed to be sterilized by a controlled ethylenoxide gasing or γ-irradiation.

5.5 Terracotta Army in Xian, China (polychrome coatings)

In order to improve the conservation techniques for the preservation and protection of the "oriental lacquer" coating the terracotta warriors in the mausoleum of the first chinese emperor Qin Shi Huang in Lintong, the mechanisms of alteration of the polychrome coatings and the subsequent conservation treatments were analysed under the coordination of the Bavarian Institute for Preservation of Historic Buildings and Monuments in Munich (Germany) and the Museum of Terracotta Warriors and Horses of Qin Shi Huang in Lintong (China).

Microbial contaminations can be observed on nearly all materials inside the excavation in Lintong (painted layers, terracotta, wood, loam). The humidity ranges between 60 and 80 %, reaching at the bottom of the excavation site partly the dew point. In addition, it was necessary to analyse the microbial contamination of the recovered terracotta fragments and in the excavation and determine its taxonomical composition, distribution and metabolic activity (e.g. impression plates, quantification of air-borne microorganisms, ATP-analysis), to evaluate the supporting microbial growth conditions (e.g. monitoring of climate data) and to develop effective countermeasures via specific climate controls and subsequent biocidal treatments (e.g. arrangement of testfields).

In the case of the conservation procedure of the polychrome coatings on the terracotta statues, the incidental decomposition of the purely organic priming coat plays an important role. The coating is extremely sensitive to changes of its moisture content and shows extreme shrinking and deformation in dry conditions, leading to a steady loss of the historic paints. The conservation procedure starts therefore with the reduction of the dry shrinking and the consolidation of the paint layers; simultaneously, during this procedure, the microbial contamination of the fragments has to be controlled at the humidity levels (90-95 % r.H.) necessary for the preservation of the non-fixed coatings. In this context it was even necessary to test the consequences of the proposed conservation treatments on the microbial contamination in order to minimize the microbial damage of the fragments by an adequate selection of protectives.

The results of these investigations revealed that fungi are the most important contaminants of the analysed fragments and in the excavation itself; especially in the soil samples the presence of actinomycetes could be proved and from the terracotta *in situ* various cyanobacteria could be isolated. The isolated microflora possess a strong biocorrosive activity, including acid production and manganese oxidation properties. Over and above that, the microbial contamination was subjected to cause health problems within the excavation fields.

The regular application of organic biocides has to be evaluated critically. The clayish, loamy soil absorbs and neutralizes the active substances of the biocides very rapidly in its clay particles and provides during microbial mineralization an important nutritive source for the reocurring microflora.

While the function of the lacquer layer as potential nutrient source could not be proven so far, the underlying microflora tends to infiltrate and detach the paint layer from the fragments. This hidden contamination represents an important problem especially for the preservation of the oriental lacquer layer on the Terracotta Warriors.

In order to control the biodeterioration problems on the terracotta fragments and in the excavation in Lintong, the provisional recommendations were a regular climate control and ventilation in the excavation area, a controlled drying of lacquer layers, and the disinfection and subsequent biocidal treatment of the loamy soil of the bulwark with medical alcohol and an inorganic biocidal solution (e.g. 5-10 % borax in tap water). Further biocidal additives are still going to be tested. The cleaning, impregnation and consolidation of the terracotta fragments showed a considerable reduction of the microbial contamination and biodeterioration processes. Organic biocidal treatments (e.g. 0.5 % CMK in iso-propanol) were in certain cases advised during the consolidation of the sensitive lacquer layers kept under high humidity conditions, whereas in most cases the application of medical alcohol was quite sufficient to control the infecting microflora during the conservation process.

6. PROSPECTIVE NEEDS FOR AN INTERDISCIPLINARY APPROACH IN CONSERVATION MICROBIOLOGY

The punctual detection and evaluation of microbial impacts in the deterioration of cultural artifacts as well as their potential consequences for human health requires a strong interdisciplinary analytical approach.

The importance of a thorough microbiological analysis in biodeterioration studies of materials in the restoration and conservation of cultural artifacts depends mainly from the timely recognition and evaluation of microbial influenced material damages and their hygienic relevance. In this context, an optimization of a non-destructive detection and analysis of biodeterioration processes is furthermore needed.

Basical measures in a practice-related conservation of the cultural heritage will start with the control of environmental parameters favouring microbial infections and growth. Moreover, the conservation practice is expected to provide a systematic recording of its techniques, materials and treatments in order to work out guidelines to enhance or even control biodeterioration processes on cultural artifacts. Over and above that biotechnological processes will be developed using biogenic desalination and carbonation capabilities in the conservation practice.

In the long-term the protection of the materials will be achieved by the selection of microbial resistant materials and protectives, minimizing the application of microbicides also taking into account ecotoxicological considerations. Here, more efforts are demanded in the development and application of synergistic acting biocidal combinations.

ACKNOWLEDGEMENTS

The work presented here was partly supported by grants from the German Ministery of Education and Research (BMBF).

REFERENCES

Adan, O.C.G. 1994. PhD-Thesis *On the fungal defacement of interior finishes.* TU Eindhoven (Netherlands).

Averdieck, B., Deininger, Ch., Engelhart, S., Missel, Th., Philipp, W., Riege, F.G., Schicht, B. and Simon, R. 1997. Bestimmung der Konzentration biologischer Arbeitsstoffe in der Luft am Arbeitsplatz (Erster Ringversuch "Schimmelpilze"). Gefahrstoffe - Reinhaltung der Luft **57:** 129-136.

Becker, T.W., Krumbein, W.E., Warscheid, Th. and Resende, M.A.1994. Investigations into Microbiology. *In* IDEAS - Investigations into Devices against Environmental Attack on Stones, Final report H.K. Bianchi (ed.), GKSS-Forschungszentrum, Geesthacht (Germany) p. 147-190.

Flannigan, B. 1997. Air sampling for fungi in indoor environments. J. Aerosol Sci **28:** 381-392.

Flemming, H.-C. and Schaule, G. 1994. Biofouling. Werkstoffe und Korrosion **45:** 29-39.

Gallo, F. 1993. Aerobiological research and problems in libraries. Aerobiologia 9: 117-130

Hödl, I. 1994a. Selbstschutz für Archivmitarbeiter. *In* Restauratorenblätter IIC (International Institute for conservation and restoration of artistic arts) p. 73-79.

Hödl, I. 1994b. Restaurierung und Konservierung von mikroorganismen-befallenen Archivalien im Steiermärkischen Landesarchiv. *In* Restauratorenblätter IIC (International Institute for conservation and restoration of artistic arts) p. 65-72.

Koestler, R.J., Brimblecombe, P., Camuffo, D., Ginell, W.S., Graedel, T.E., Leavengood, P., Petushkova, J., Steiger, M., Urzi, C., Vergès-Belmin, V. and Warscheid, Th. 1992. How do External Environmental Factors Accelerate Change ? *In* Durability and Change - The Science, Responsibility and Cost of Sustaining Cultural Heritage, W.E. Krumbein, P. Brimblecombe, D.E. Cosgrove and S. Staniforth (eds.), John Wiley and Sons, New York p. 149-163.

Koestler, R.J., Warscheid, Th. and Nieto, F. 1997. Biodeterioration: Risk Factors and their Management. *In* Saving our Cultural Heritage: The Conservation of Historic Stone Structures, R. Snethlage and N.S. Baer (eds.), John Wiley and Sons, New York p. 25-36.

Koestler, R.J. 1999. Polymers and resins as food for microbes. *In* Of Microbes and Art: The Role of Microbial Communities in the Degradation and Protection of Cultural Heritage, O. Ciferri, P. Tiano and G. Mastromei (eds.), Plenum p. 153-167.

Kumar, R. and Kumar A.V. 1999. Biodeterioration of stone in tropical environments - An overview. Research in Conservation Series, The Getty Conservation Institute, Los Angeles.

Leznicka, St. 1992. Antimicrobial protection of stone monuments with p-hydroxybenzoic acid esters and silicone resin. *In* Proceedings of the VIIth International Congress on Detrioration and Conservation of Stone J.D. Rodrigues, F. Henriques and F.T. Jeremias (eds.), Laboratorio Nacional de Engenharia Civil, Lisbon (Portugal) p. 481-490.

May, E. and Lewis, F.J. 1988. Strategies and techniques for the study of bacterial populations on decaying stonework. *In* Proceedings of the VIth International Congress on Deterioration and Conservation of Stone Nicolaus Copernicus University (ed.), Torun (Poland) p. 59-70.

May, E., Lewis, F. J., Pereira, S., Tayler, S., Seaward, M.R.D. and Allsopp, D. 1993. Microbial deterioration of building stone - a review. Biodeterioration abstracts **7/2:** 109-123.

Mitchell, R. and Ji-Dong Gu 1999. Interactions between air pollutants and biofilms on historic limestone. Of Microbes and Art: The Role of Microbial Communities in the Degradation and Protection of Cultural Heritage. International Conference on Microbiology and Conservation, Florence, June 1999 p.143-145.

Nugari, M.P. 1999. Interference of antimicrobial agents on stone. Of Microbes and Art: The Role of Microbial Communities in the Degradation and Protection of Cultural Heritage. International Conference on Microbiology and Conservation, Florence, June 1999 p. 211-214.

von Plehwe-Leisen, E., Warscheid, Th. and Leisen, H. 1996. Studies of longterm behaviour of conservation agents and of microbiological contamination on twenty years exposed treated sandstone cubes. *In* Proceedings of the 8th international congress on deterioration and conservation of stone Vol. 2, Riederer, J. (ed.), Rathgen-Forschungslabor, Berlin (Germany) p. 1029 - 1037.

Richardson, B. A. 1988. Control of microbial growth on stone and concrete. *In* Biodeterioration 7, D.R. Houghton, R.N. Smith and H.O.W. Eggins(eds.), Elsevier Applied Science, London and New York p. 101-106.

Saiz-Jimenez, C. 1995. Deposition of air-borne organic pollutants on historical buildings. Atmospheric Environment **27**: 77-85.

Schwarz, H.J. 1996. Die gotischen Wandmalereien der Allerheiligenkapelle im Kloster Wienhausen: Untersuchungen zur Erstellung eines Restaurierungs- und Erhaltungskonzeptes. Berichte zur Denkmalpflege in Niedersachsen **4**: 140-143.

Tiano, P., Camaiti, M. and Accolla, P. 1995. Methods for evaluation of products against algal biocoenosis of monumental fountains. *In* Methods of evaluating products for building materials in monuments, ICCROM-International Colloquium, Rome p. 75-86.

Tiano, P., Biagiotti, L. and Bracci, S. 1999. Biodegradability of products used in monuments conservation. *In* Proceedings of the ICMC "Of Microbes and Art" - The role of Microbial Communities in the degradation and protection of cultural heritage, O. Ciferri (ed.) p. 197-199.

Wakefield, R.D. and Jones, M.S 1996. Some effects of masonry biocides on intact and decayed stones. *In* Proceedings of the 8th international congress on deterioration and conservation of stone Vol. 2, Riederer, J. (ed.), Rathgen-Forschungslabor, Berlin (Germany) p.703-716.

Warscheid, Th., Petersen, K. and Krumbein, W. E. 1988. Effects of cleaning on the distribution of microorganisms on rock surfaces. *In* Biodeterioration 7, Houghton, D. R., Smith, R. N. and Eggins, H. O. W. (eds.), Elsevier Applied Science, London p. 455-460.

Warscheid, Th., Oelting, M. and Krumbein, W.E. 1991. Physico-chemical aspects of biodeterioration processes on rocks with special regard to organic pollution. Int. Biodet. **28**: 37-48.

Warscheid, Th. and Krumbein, W.E. 1996. Biodeterioration of Inorganic Nonmetallic Materials - General Aspects and Selected Cases. *In* Microbially Induced Corrosion of Materials, Heitz / Sand / Flemming (eds.), Springer-Verlag p. 273-295.

Warscheid, Th., Becker, T.W., Braams, J., Brüggerhoff, S., Gehrmann, C., Krumbein, W.E. and Petersen 1993. Studies on the temporal development of microbial infection of different types of sedimentary rocks and its effect on the alteration of the physico-chemical properties in building materials. *In* Proceedings of the International RILEM/UNESCO Congress "Conservation of Stone and Other Materials" Vol. 1: Causes of Disorders and Diagnosis, M.-J. Thiel (ed.), RILEM, Paris p.303-310.

Warscheid, Th. 1996a. Impacts of microbial biofilms in the deterioration of inorganic building materials and their relevance for the conservation practice. Internationale Zeitschrift für Bauinstandsetzen **2**: 493-504.

Warscheid, Th. 1996b. Biodeterioration of Stones: Analysis, Quantification and Evaluation. *In* Proceedings of the 10th International Biodeterioration and Biodegradation Symposium, Dechema-Monograph 133, Frankfurt p. 115-120.

Warscheid, Th. and Kuroczkin, J. 1997, Preventive and Protective Treatments against Biodeterioration of Stone. *In* Proceedings of the Residential Course "Biotechnology and the preservation of cultural artifacts", Biotechnology Foundation (ed.), Turin.

Warscheid, Th. and Kuroczkin, J. 1999. Biodeterioration of Stones. *In* Studies in Museology, Biodeterioration of cultural properties, R. J. Koestler and A.E. Charola (eds.), Butterworth-Heinemann, submitted.

Wendler, E. 1997. New materials and approaches for the conservation of stone. *In* Saving our Cultural Heritage: The Conservation of Historic Stone Structures, R. Snethlage and N.S. Baer (eds.), John Wiley and Sons, New York p. 181-196.

BACTERIAL CARBONATOGENESIS AND APPLICATIONS TO PRESERVATION AND RESTORATION OF HISTORIC PROPERTY

Sabine Castanier[1], Gaële Le Métayer-Levrel[1], Geneviève Orial[2], Jean-François Loubière[3] and Jean-Pierre Perthuisot[1]

[1]Laboratoire de Biogéologie et Microbiogéologie, Université de Nantes, 2 rue de la Houssinière, F-44072 Nantes cedex 03, France; [2]Laboratoire de Recherche des Monuments Historiques, 29 rue de Paris, F-77420 Champs sur Marne, France; [3]Société CALCITE S.A., 43 rue Jules Guesde, F-92300 Levallois-Perret, France.

Key words: biomineralisation, bacteria, historic property

Abstract: The biomineralisation process is based on the ability of certain bacteria to produce solid Ca-carbonate. The scientific background is first presented, as far as aerobiotic metabolic pathways, biological processes and solid products are concerned. The process which consists in letting bacteria produce a carbonate protecting scale (biocalcin) on treated surfaces was developed through laboratory and life-size experiments. It is to date applied on buildings at the industrial scale. The process was also applied to statuary. Lastly, further microbiotechnical developments for restoration of limestone works of art are presented.

1. INTRODUCTION

Bacterial carbonatogenesis has been suspected since the beginning of present century. Quite recently, after number of observations and laboratory experiments, this idea has been accepted by part of the geologists' community. Meanwhile, it nowadays gives way to technical and industrial applications notably for preservation and restoration of buildings and historic patrimony.

2. SCIENTIFIC BACKGROUND

In nature carbonate precipitation may theoretically occur following several known processes : i. abiotic chemical precipitation from saturated solutions by evaporation, temperature increase and/or pressure decrease; ii. external or internal skeleton production by eukaryotes; iii. lowering of CO_2 pressure under effect of autotrophic processes (photosynthesis, methanogenesis...); iv. fungal mediation (Callot et al., 1985; Verrecchia and Loisy, 1997); v. heterotrophic bacterial mediation. As a matter of fact, bacterial contribution to limestone formation has been suspected for years (Drew 1910a; 1910b; Berkeley, 1919; Kellerman, 1915; Lipmann, 1924; Molish, 1924; Nadson, 1928; Krumbein, 1968, 1974, 1978) but remained controversial until recent experiments in microbiogeological laboratories investigated the metabolic pathways involved, the modes and conditions of solid particles formation, and evaluated bacterial carbonate productivity (Krumbein, 1979a, 1979b; Riege et al., 1991; Castanier, 1987; Le Métayer-Levrel, 1996; Castanier et al., 1997, 1998).

2.1 The metabolic pathways of bacterial Ca carbonate formation

The production of Ca-carbonate particles through bacterial mediation follows different ways.

2.1.1 Autotrophic pathways

In autotrophy, three metabolic pathways are involved: non-methylotrophic methanogenesis (Marty, 1983), anoxygenic photosynthesis and oxygenic photosynthesis. All three pathways use CO_2 as carbon source to produce organic matter. Thus, they induce CO_2 depletion of the medium or of the immediate environment of the bacteria. When calcium ions are present in the medium, such a depletion favours calcium carbonate precipitation. We are not discussing this question any further since such processes cannot be used for stone protection.

2.1.2 Heterotrophic pathways

In heterotrophy two bacterial processes may occur, often concurrently.

Passive precipitation or passive carbonatogenesis operates by producing carbonate and bicarbonate ions and inducing various chemical modifications in the medium that lead to the precipitation of calcium carbonate. Two metabolic cycles can be involved : the nitrogen cycle and the sulphur cycle.

In the nitrogen cycle, passive bacterial precipitation follows three different pathways: i. the ammonification of amino-acids in aerobiosis, in the presence of organic matter and calcium. ii. the dissimilatory reduction of nitrate in anaerobiosis or microaerophily, in the presence of organic matter and calcium and iii. the degradation of urea or uric acid in aerobiosis, in the presence of organic matter and calcium.

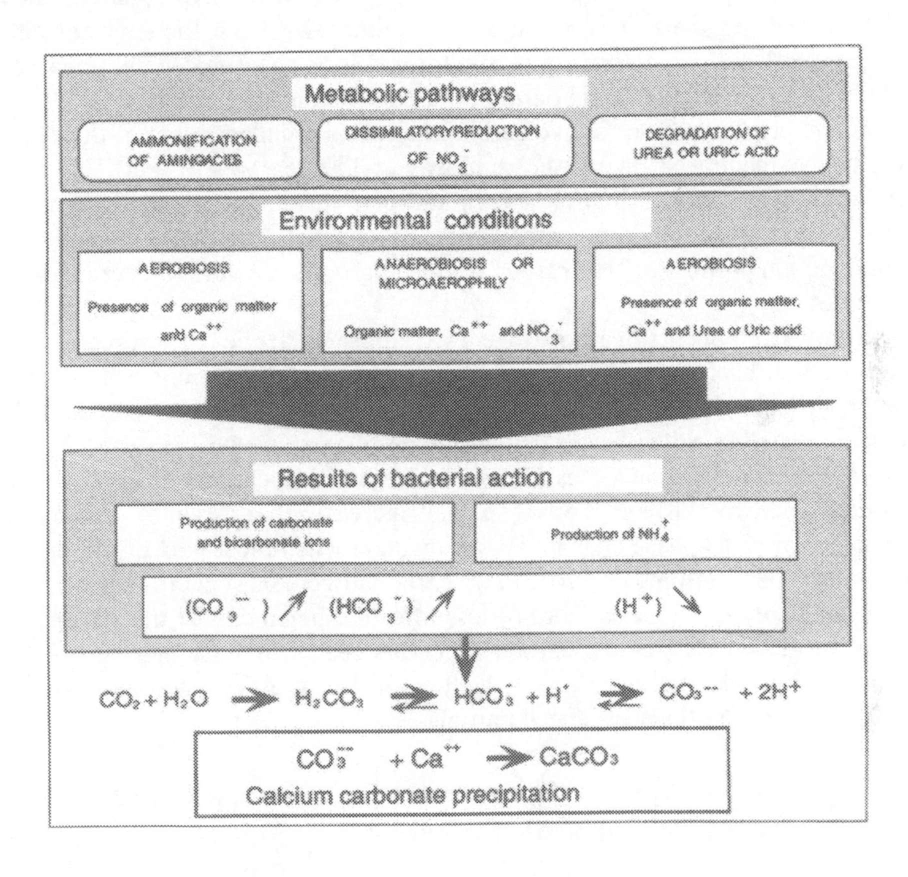

Figure 1. The passive bacterial precipitation of carbonates in the nitrogen cycle.

Both urea and uric acid result from eukaryotic activity, notably that of vertebrates. These three pathways induce production of carbonate and bicarbonate ions and, as a metabolic end-product, ammonia, which causes pH increase (Fig. 1). When the H^+ concentration decreases, the carbonate-bicarbonate equilibria are shifted towards the production of CO_3^{--} ions. If calcium ions are present, calcium carbonate precipitation occurs. If Ca^{++} (and/or other divalent cations) are lacking in the medium, carbonate and bicarbonate ions accumulate, and the pH increases and bacterial activity may favour zeolite formation. This happens in soda lakes, e.g. in Kenya (Castanier et al., 1993).

In the sulphur cycle, bacteria use a single metabolic pathway : the dissimilatory reduction of sulphate. The environment must be anoxic, and rich in organic matter, calcium and sulphate. Using this pathway, bacteria produce carbonate, bicarbonate ions and hydrogen sulphide. If calcium ions are present, the precipitation of Ca-carbonate depends on the hydrogen sulphide behaviour. When the hydrogen sulphide is withdrawn from the environment, pH increases and Ca-carbonate precipitates. Since anaerobic conditions are required, such a process could hardly be used for stone protection.

Active precipitation or active carbonatogenesis is independent of the other previously mentioned metabolic pathways. The carbonate particles are produced by ionic exchanges through the cell membrane by activation of calcium and/or magnesium ionic pumps or channels, probably coupled to carbonate ion production. Numerous bacterial groups are able to operate such processes.

In all experiments, carbonatogenesis appears to be the response of heterotrophic bacterial communities to an enrichment of the milieu in organic matter. After a phase of latency, there is an exponential increase of bacterial numbers together with the accumulation of metabolic end-products. These induce an accumulation of carbonate and hydrogenocarbonate ions in the medium and, by different ways, a pH increase that favours carbonate precipitation. This phase ends into a steady state when most part of the initial enrichment is consumed. Particulate carbonatogenesis occurs during the exponential phase and ends more or less after the beginning of the stationary phase. In most cases, active carbonatogenesis seems to start first and to be followed by the passive one which induces the growth and shape modifications of initially produced particles.

2.2 Relationships between bacteria, carbonate minerals and environmental conditions

2.2.1 In eutrophic conditions

In experiments with eutrophication, i.e. with massive inputs of organic matter, the first solid products are probably amorphous and perhaps hydrated at the beginning (Castanier et al., 1988). They appear on the surface of the bacterial bodies as patches or stripes that extend and coalesce until forming a rigid coating (cocoon). In other cases, solid particles, formed inside the cellular body, are excreted from the cell (excretates). All these tiny particles, including more or less calcified bacterial cells, assemble into biomineral aggregates which often display "precrystalline" or rather "procrystalline" structures. Sometimes the bacterial bodies themselves present an angular crystalline shape as if they would contain a growing single crystal. At this stage of evolution and in numerous cases, the calcified bacterial cells tend to arrange themselves into nearly crystalline structures and sometimes into dendritic or fibroradial fabrics (Castanier et al., 1989). The angles of such structures are generally close to but not exactly those of the rhombohedral system nor those of the orthorhombic one as if bacteria dislike the crystallographic physical rules.

The primary aggregates grow and form secondary biocrystalline assemblages or build-ups which progressively display more crystalline structures with growth. Tetrahedral assemblages and pentagonal faces are often observed. This phase should correspond to the passive carbonatogenesis. Different types of crystallogenetic sequences seem to follow the different metabolic pathways (Castanier, 1987).

Thus, after eutrophication, bacterial activity is very high at the beginning and early solid products, as well as biomineral aggregates, have poorly defined crystal structure which is overwhelmed by biological luxuriant processes. For example, such conditions are suitable for the formation of round bodies with a radial internal structure, i.e. ooids (Castanier et al., 1989).

2.2.2 In oligotrophic conditions

In oligotrophic environments, i.e. when the organic matter input is scarce, for example in karstic caves (Le Métayer-Levrel, 1996; Le Métayer-Levrel et al., 1997), only a few kinds of bacteria may live. It is often observed that bacterial carbonate material as well as organic material are rapidly incorporated into large, well-shaped crystals the growth of which should be slow. After dissolution of crystals with EDTA, SEM imagery clearly shows that the bacteria act as axes of nucleation to calcite rhombohedra. When the crystals have almost totally been removed by dissolution, there remains a network of organic filaments.

These observations and experiments show that in both nutritional conditions bacteria play a major role in crystallisation, both in supplying carbonate matter and in its physical structure (Perthuisot et al., 1997). On the contrary, in oligotrophic conditions, bacterial production rate is low so that the crystallographic rules soon overcome the biological primary disorder. Clearly, the latter conditions are not suitable for stone protection.

2.3 Productivity of heterotrophic bacterial carbonatogenesis

Quantitatively, the production of solid carbonate depends essentially upon the strains in the bacterial population, the environmental conditions (temperature, salinity, etc.), the quality and quantity of available nutrients, and time.

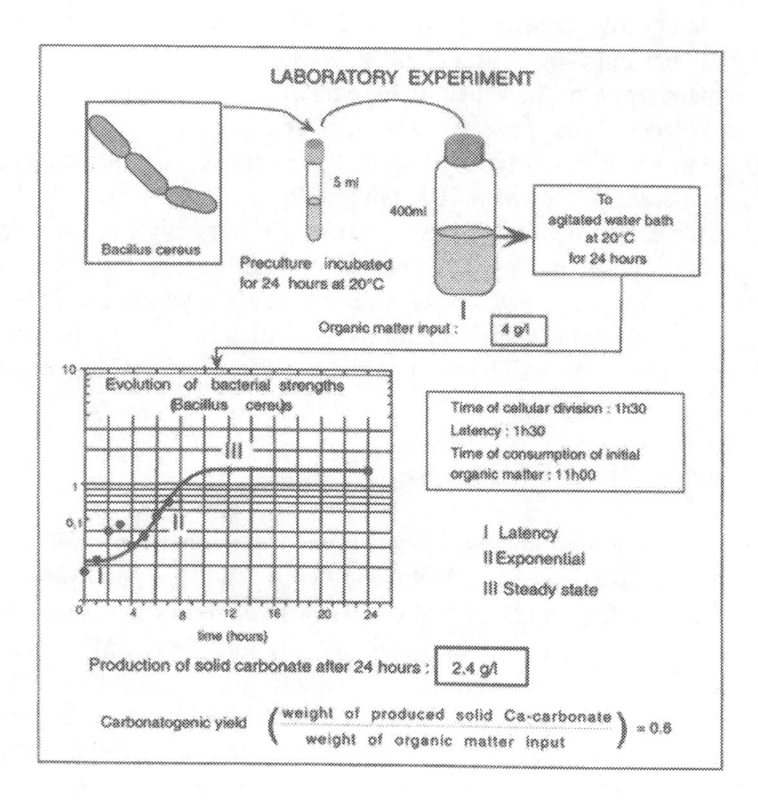

Figure 2. An example of a laboratory experiment aiming at evaluating the potentialities of bacterial carbonatogenesis

In the laboratory, all heterotrophic metabolic pathways were tested and a number of experiments were carried out with *Bacillus cereus*, which is a heterotrophic bacterium able to undertake ammonification of amino-acids and dissimilatory nitrate reduction. In the experiment presented in Fig. 2, the medium initially contains 4 g organic matter per litre (Castanier, 1984; Le Métayer-Levrel, 1996). As usual, the phase of latency is followed by an exponential increase in bacterial growth together with the accumulation in the medium of metabolic end-products: carbonate, bicarbonate and ammonia ions.

This phase is followed by the stationary phase. Ca-carbonate precipitation occurs during the exponential phase and ends more or less after the beginning of the steady state.

In the above mentioned experiment, with a nutritive input of 4 gl^{-1} of organic matter, 2.4 grams of calcite were obtained per litre per day (Le Métayer-Levrel, 1996). The "carbonatogenic yield" (or calcium-carbonate yield) may be defined as the ratio of the weight of organic matter input to the weight of calcium carbonate produced. In the case presented here it is 0.6. We presently keep in the laboratory about a hundred carbonatogenic strains collected from various natural environments. Most of them display carbonatogenic yields around 0.5, sometimes more, the lowest yield measured being 0.2. These numbers refer to experiments with monospecific cultures. In nature, carbonatogenesis is generally carried out by plurispecific populations so that sedimented organic matter may be totally mineralised into carbonates.

The environmental conditions of heterotrophic bacterial metabolic pathways are diverse (aerobiosis, anaerobiosis, microaerophily). However, carbonate precipitation always appears to be the response of heterotrophic bacterial communities to enrichment of the environment in organic matter.

3. FROM BACTERIAL CARBONATOGENESIS TO PROTECTION AND REGENERATION OF LIMESTONES IN BUILDINGS AND HISTORIC PATRIMONY

In natural environments, rocks, and, notably, limestones deteriorate under the effect of weathering. Furthermore, after these geological materials are quarried out and used in building they are submitted to new types of alteration. Atmospheric water collects environmental pollutants and superficial moisture allows these to penetrate into the pores of the stone, adding to natural processes. Concurrently, bacterial acidification may occur and be very harmful, especially to limestone (Rautureau, 1991; Phillipon et al., 1992).

In order to stop further deterioration of degraded buildings, particularly those classified as historic monuments, currently employed preventive or curative treatments prescribe the use of synthetic resins (silane, polysiloxane) that polymerize inside the stones pores. Such treatments, based on chemical products, tend to produce a superficial skin which deteriorates with age, tends to peel off, and thus requires constant maintenance (Camaiti et al., 1988). Additionally, large quantities of organic solvents are used and contribute to pollution. The ability of bacteria to promote the precipitation of carbonates led to the proposition of a new kind of treatment (Boquet et al., 1973;

Adolphe and Billy, 1974; Adolphe, 1981). A patent was taken out (Adolphe et al., 1990) and a society (Calcite S.A.) was created.

The industrial optimization of the process and of potential further developments was devolved to the Laboratory of Biogeology and Microbiogeology of Nantes University. Additionally, the adaptation of these new technologies to the conservation of historic monuments has been and still is studied in collaboration with the Historic Monuments Research Laboratory.

3.1 The protection of buildings façades - The biomineralisation process

Several sorts of limestones are used for building and sculpture. Among these, two categories may be distinguished according to the size of their constituents and their porosity. The first category includes fine-grained limestones, which may have high porosity but be formed of small (less than 10 μm) pores. This is the case of the famous Tuffeau limestone which is a kind of siliceous chalk used widely in historic monuments in Western France. The second category is composed of generally biodetrital limestones of variable porosity formed of both larger grains and larger pores (more than 10 μm wide), such as the Saint-Maximin limestone and the Saint-Vaast limestone which were also used for experiments (Le Métayer-Levrel, 1993, 1996).

The biomineralisation process consists of first spraying the entire surface to be protected with a suitable bacterial suspension culture (Fig. 3). Afterward, the deposited culture is fed daily or every two days with the suitable medium in order to create a surficial calcareous coating scale, the "biocalcin". The nutritional medium is designed to stimulate the active and passive bacterial production of carbonate through the nitrogen cycle metabolic pathways which are the only pathways to be activated in operational conditions i.e. in aerobiosis and microaerophily (Castanier, 1987; Castanier et al., 1997, 1998). Usual industrial and financial constraints restrict the number of feeding applications to five, but treatment of Historic Patrimony may be less limited.

3.1.1 Preliminary experiments in the laboratory

Initially, several scientific investigations were undertaken. First, it was necessary to collect bacterial communities from natural carbonate producing environments. Various environments were sampled including karstic deposits, lagunas of the Bolivian Altiplano and Pamukkale calcareous falls. Carbonatogenic strains were isolated, tested for their carbonatogenic yield and stored in the laboratory. Most species were identified at the Pasteur Institute. To date the collection contains more than 100 strains. Secondly, it was necessary to test various culture media in order to see if they met industrial financial constraints. Notably, it was necessary to replace the expensive components of media currently used in the laboratory with cheaper ones.

The process has been first simulated on miniature walls composed of both categories of the limestones mentioned before. Since limestones may be composed at least partly of fossilized bacterial bodies, it is often difficult to distinguish the calcin crust formed after treatment from that already present. Therefore an inert non-limestone support was necessary to act as control and vouch for the validity of the test. Thus another test was carried out on heat-resistant bricks. Several carbonatogenic bacterial strains have been tested as well as two feeding frequencies (once every 24 hours and once every 48 hours). The daily frequency is more suitable for fine-grained limestone; the other one for coarse-grained limestone. The best operating strains differ depending on the type of limestone. In the previously mentioned conditions, carbonate producing bacteria colonize the entire surface of the stone (including the cristobalite spheres in the Tuffeau limestone), which is rapidly coated by the "biocalcin". This coating forms a smooth blanket, several micrometers thick, composed mainly of encrusted bacterial bodies mixed with carbonate excretions. These are mineral particles formed inside the bacterial cell and expelled afterwards (Castanier, 1987; Castanier et al., 1997, 1998). It partially fills the voids formed by pores at the surface and thus becomes rooted in the structure of the stone. The biocalcin insures the protection of limestone by restricting exchange between the interior of the rock inside and external atmosphere and, additionally, by limiting the penetration of degrading agents into the stone.

3.1.2 Life-size experiments

The first life-size experimentation was carried out in Thouars (Deux Sévres) on the SE tower of Saint Médard Church. This church was built during the 12th century with Tuffeau limestone and, since that date, has been restored several times. The tower was entirely restored in 1988 so that the surface of the Tuffeau processed by biomineralisation was only slightly damaged before treatment. The treatment was applied in June 1993 on an area of 50 m^2. The evolution of the bacterially-produced biocalcin while exposed to weather variations was evaluated twice, first after 6 months and again after one year. Six observations and tests were performed in 20 points over the treated surface: macroscopic observation, SEM imagery, measurement of surficial permeability (by measuring the time of water absorption using a water pipe), determination of bacterial population, evaluation of surface roughness by imprint molding, and colorimetry. The last measurement, made in July 1997, confirmed the good quality and consistency of the biocalcin. Furthermore, the abundant development of carbonatogenic bacterial populations prevented the development of autochtonous acidifying bacterial consortia. To date, six years after treatment, the external aspect of the tower has not changed.

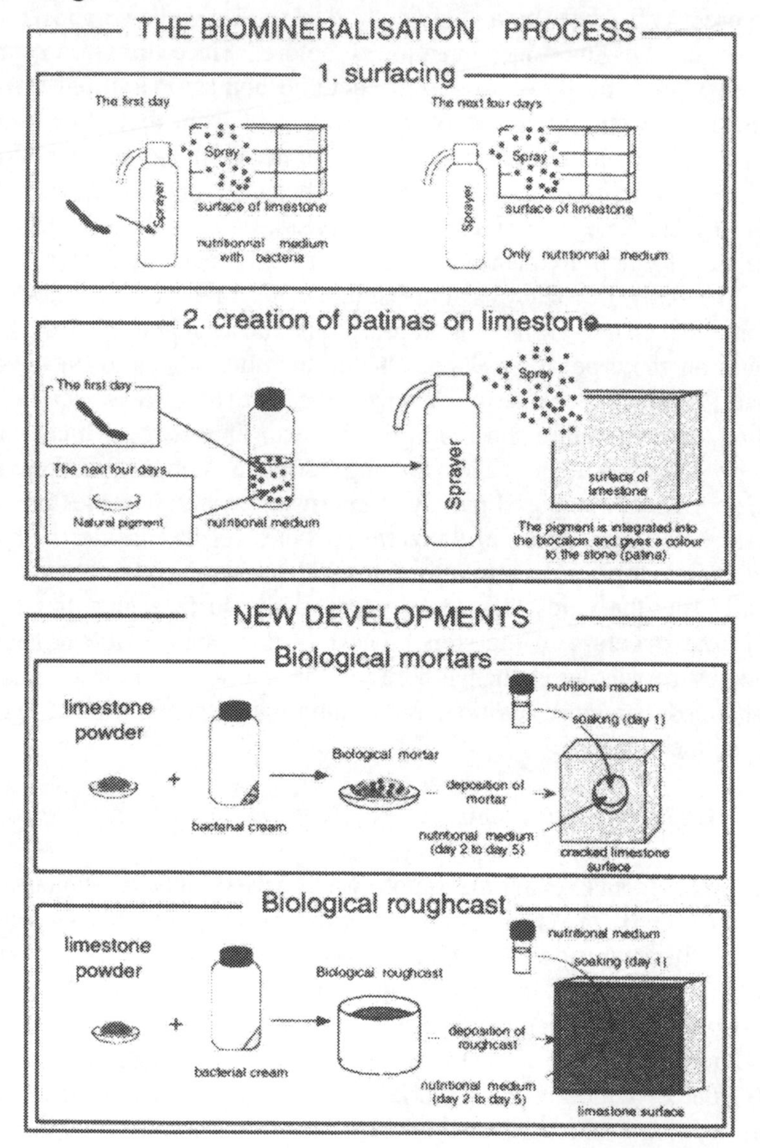

Figure 3. Technologies developed from the bacterial carbonatogenesis concept

These experiments validated the viability and illustrated the value of the biomineralisation process on an upright wall exposed to natural weathering processes. The biocalcin provides the stone with a protective surface that reduces its absorption of water (with a 1/5 ratio) yet retains its permeability to gas, without detracting from its aesthetic appearance.

3.1.3 Large scale developments

On the northern façade of Saint Médard church, two portions each of 200 m^2 in surface were treated in 1997 and 1998. Since they are parts of a historical monument, they will be surveyed by the same methods mentioned above

In addition, the façades of several private buildings in Paris have already been industrially treated by the biomineralisation process. Among them, the Plaza Hotel and the Marks and Spencer's store may be mentioned.

3.2 Applications to the statuary

The same surficial biomineralisation process was also experimentally applied to limestone statuary (Le Métayer-Levrel, 1996). Twenty prototype statues comprising all usual sculptural features were manufactured for a test program aiming at studying the ageing of biomineralisation protection coating on statues placed in different climatic environments. Both Tuffeau and Saint Maximin limestones were studied under five different environmental conditions (Fig. 4).

Three outdoor sites for ageing experiments were chosen: rural (Loire Atlantique), urban (Paris), and littoral (Vendée) environments. Another set of statues was placed in an indoor protected environment (Historic Monuments Research Laboratory). The last set was placed in an ageing acceleration climatic station. The surfacing biomineralisation process was applied to statues in January 1994 in the same manner as described above. Pairs of treated and untreated statues were installed after the treatment had matured for one year in the laboratory. Series of measurements and observations have been performed at regular intervals. After 15 months exposure, they revealed that the biocalcin has undergone a similar evolution in all outdoor sites. As compared to untreated statues, treated ones displayed little degradation and then only in parts where rainwater tended to accumulate. The accelerated ageing gave remarkable results. Untreated prototype statues were heavily degraded whereas treated ones remained intact. After 3 years exposure in all outdoor situations and on each kind of limestone, the biocalcin was scarcely degraded and still covering processed statues.

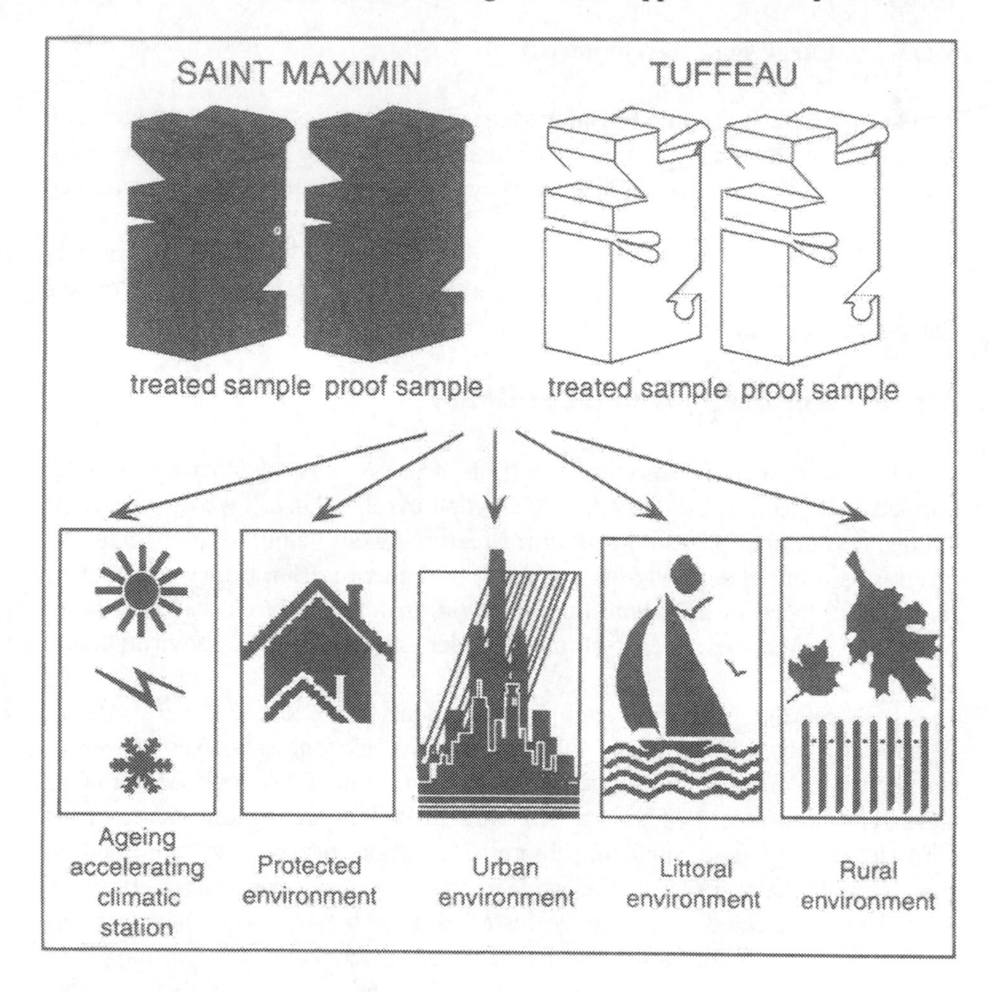

Figure 4. Sketch plan of experiments on statuary

The last group of measurements and observations were performed in spring 1999 i.e. after 4 years exposure. They show that biocalcin does not display any change in the indoor protected environment. On the contrary, erosion of the biocalcin at the surface of the outdoors statues has more or less progressed depending upon the stone and the environment. Erosion is globally more obvious on the Tuffeau since this stone contains cristobalite spheres. These progressively emerge from the biocalcin and are eroded because, on top of them, the biocalcin is probably thinner and, possibly, poorly rooted. The worse environment appears to be the rural one due to steady and heavy agricultural pollution by nitrates and ammonia. The stone is covered with cyanobacteria, algae and fungi and the biocalcin heavily degraded. The littoral environments is not far better and while gypsum or salt have not been observed, the treated stones are colonised by fungi and lichens. Curiously, in the urban site (Paris), the statues behave better. The biocalcin is scarcely degraded and only fungi are present.

It is worth noting that, in all exposure sites, processed statues kept a better look than untreated ones. Besides, in the worse degradation cases, the biocalcin still infills the pores of the rock surface and thus keeps protecting the stone.

Thus, surficial biomineralisation by carbonatogenic bacteria also appears as a good protecting process for limestone statuary. This program is ongoing.

3.3 Creation of patinas

When applying the biomineralisation process on limestone façades it is also possible to create a surficial patina by adding natural pigments to the nutritional medium (Fig. 3). The pigments are integrated into the biocalcin and thus give a persistent light colouring to the stone. Of course, before processing, experiments are necessary to select the appropriate pigment and the correct concentration needed to obtain a specific tone. Such a technique may be used, for example, to conceal a few replaced stones on a monument façade.

4. FURTHER TECHNICAL DEVELOPMENTS OF THE BACTERIAL CARBONATOGENESIS CONCEPT

Another application of bacterial carbonatogenesis is the conception of biological mortars or cements (Castanier *et al.*, 1995). They are made of a mixture of bacteria, finely ground limestone and nutritional medium in variable proportions (Fig. 3). They can be used to attach small pieces broken out of statues or to fill small cavities on limestone surfaces. It is also possible to include pigments in the mixture in order to obtain coloured biological mortars. This is a new concept that may be very useful in limestone monument restoration as it rapidly recreates a material similar to the stone as well as avoids the formation of discontinuities and subsequent breaks. It has been successfully tested on a small scale on sculptures of Amiens Cathedral and will be soon tested on a larger scale in the restoration of the porch of the church of Argenton-Château (Deux Sèvres). A program of experiments in the laboratory is ongoing. It aims at determining the best bacterial and mineral components to be used, e. g. suspension or floc.

5. CONCLUSION

The biomineralisation process has brought a new element into the field of stone and monument restoration. As far as the protection of façades is concerned, at the moment the process appears to be most effective on clean and healthy stone and thus should be applied in the future to previously cleaned stone. However, pre-cleaning should exclude any chemical product which could inhibit the development of bacteria. For example the photonic desincrustation using a Laser beam, which was developed by the Historic Monuments Research Laboratory, would be the ideal pre-cleaning method to use in conjunction with the biomineralisation process (Orial et al., 1993).

Finally, a large field of study is open in which the biomineralisation process will be applied to statues as a means of protecting stone surfaces. Besides, bacterial carbonatogenesis leads to possible means of restoration such as consolidation and adhesion of broken pieces.

Compared to the more traditional methods, the techniques described here differ mainly in their natural and ecological character and in the fact that the protection acquired tends to increase with age. All of the applications mentioned here tend to recreate a naturally produced material in a way that is similar to the limestone substrate formation. As a matter of fact, these applications follow the same metabolic pathways that result in the formation of limestone in nature.

REFERENCES

Adolphe, J.-P. 1981. Observations et experimentations géomicrobiologiques et physicochimiques des concrétionnements carbonatés continentaux actuels et fossiles. Doctorate Thesis, University P. & M. Curie, Paris.

Adolphe, J.-P. and Billy, C. 1974. Biosynthèse de calcite par une association bactérienne aérobie. C. R. Acad. Sc. **278**: 2873-2875.

Adolphe, J.-P., Loubière, J.-F., Paradas, J. and Soleilhavoup, F. 1990. Procédé de traitement biologique d'une surface artificielle. European patent n° 90400G97.0 (after French patent n° 8903517, 1989).

Berkeley C. 1919. A study of marine bacteria. Straits of Georgia B.C. Proc. Trans. R. Soc. Can., Ottawa section **5**, **13**: 15-43.

Boquet, E., Boronat, A. and Ramos-Cormenzana, A. 1973. Production of calcite (calcium carbonate) crystals by soil bacteria is a common phenomenon. Nature **246**: 527-529.

Callot G., Guyon A. and Mousain, D. 1985. Inter-relations entre aiguilles de calcite et hyphes mycéliens. Agronomie **5**: 209-216.

Camaiti, M., Borselli, G. and Matteoli, U. 1988. Prodotti consolidanti impiegati nelle operazioni di restauro. Edilizia **10**: 125-134.

Castanier, S. 1984. Étude de l'évolution quantitative et qualitative des populations bactérienne précipitant le carbonate dans différents cas artificiels de confinement réalisés à partir d'eau et de sédiment lagunaires méditerranéens. Thesis, University of Aix-Marseille II, Marseille, France.

Castanier, S. 1987. Microbiogéologie : Processus et modalités de la carbonatogenèse bactérienne. Doctorate thesis, University of Nantes, France.

Castanier, S., Maurin, A. and Perthuisot, J.-P. 1988. Les Cugnites : carbonates amorphes de Ca et Mg, précurseurs possibles de la dolomite. C. R. Acad. Sci. Paris **306**: 1231-1235.

Castanier, S., Maurin, A. and Perthuisot J.-P. 1989. Production microbienne expérimentale de corpuscules carbonatés sphéroïdaux à structure fibro-radiaire. Réflexions sur la définition des ooïdes. Bull. Soc. Géol. Fr. **5**: 589-95.

Castanier, S., Bernet-Rollande, M.-C., Maurin, A. and Perthuisot, J.-P. 1993. Effects of microbial activity on the hydrochemistry and sedimentology of Lake Logipi, Kenya. Hydrobiologia **267**: 99-112.

Castanier S., Le Metayer-Levrel G. and Loubière, J.-F. 1995. Nouvelles compositions pour mortier biologique, procédé de recouvrement d'un surface ou de comblement d'une cavité à l'aide des compositions. French Patent n° 95 05861.

Castanier, S., Le Métayer-Levrel, G. and Perthuisot, J.-P., 1997, La carbonatogenèse bactérienne. *In* Hydrologie et géochimie isotopique F. Causse and F. Gasse (eds.), ORSTOM, Paris, pp. 197-218.

Castanier, S., Le Métayer-Levrel, G. and Perthuisot, J.-P. 1998. Bacterial roles in the precipitation of carbonate minerals. *In* Microbial Sediments R. Riding and S. Awramik, (eds.), Springer Verlag, Berlin (in press).

Drew, G.H. 1910a. The action of some denitrifying bacteria in tropical and temperate seas, and the bacterial precipitation of calcium carbonate in the sea. J. Mar. Biol. Assoc. **IX**: 142-155.

Drew, G.H. 1910b. On the precipitation of calcium carbonate in the sea by marine bacteria, and on the action of denitrifying bacteria in tropical and temperate seas. J. Mar. Biol. Assoc. **IX**: 479-523.

Kellerman, K.F. 1915. Relation of bacteria to deposition of calcium carbonate. Geol. Soc. Amer. Bull. **26**: 58

Krumbein, W.E. 1968. Geomicrobiology and geochemistry of lime crusts in Israel. *In* Recent developments in carbonate sedimentology in Central Europe, G. Muller and G.M. Friedman, (eds.), Springer, Berlin, Heidelberg, New York, pp. 134-147.

Krumbein, W.E. 1974. On the precipitation of aragonite on the surface of marine bacteria. Naturwiss. **61**:167-177.

Krumbein, W.E. 1978. Algal mats and their lithification. *In* Environmental biogeochemistry and geomicrobiology, W.E. Krumbein (ed.), The Aquatic Environment, Ann Arbor Science Publ. Inc., Michigan, **1** pp. 209-225.

Krumbein, W.E. 1979a. Calcification by bacteria and algae. *In* Biogeochemical Cycling of Mineral-Forming Elements, P.A. Trudinger and D.J. Swaine (eds.), Elsevier, Amsterdam, pp. 47-68.

Krumbein, W.E. 1979b. Photolithotrophic and chemoorganotrophic activity of bacteria and algae as related to beachrock formation and degradation (Gulf of Aqaba, Sinai). Geomicrobiol. J. **1**: 139-203.

Le Métayer-Levrel G. 1993. Biominéralisation de surfaces : application à la protection des pierres de taille. Diploma of University Research, University of Nantes, France.

Le Métayer-Levrel, G. 1996. Microbiogéologie du carbonate de calcium. Applications industrielles. Implications géologiques. Doctorate thesis, University of Nantes, France.

Le Métayer-Levrel, G., Castanier, C., Loubière, J.-F. and Perthuisot, J.-P. 1997. La carbonatogenèse bactérienne dans les grottes. Étude au MEB d'une hélictite de Clamouse, Hérault, France. C. R. Acad. Sci. Paris, **325**: 179-184.

Lipmann, C.B. 1924. Further studies on marine bacteria with special reference to the Drew hypothesis on $CaCO_3$ precipitation in the sea. Carnegie Inst. Washington Publ. **391**, 26: 231-248.

Marty, D. 1983. Cellulolyse et méthanogenèse dans les sédiments marins. Doctorate thesis University of Aix-Marseille I, Marseille, France.

Molish, H. 1924. Uber kalkbacterien und ausere kalkfallende pilze. Zentralblat Bakteriol. II **65**:130-139.

Nadson, G.A. 1928. Beitrag zur Kenntis der baketriogen Kalkabla gerungen. Arch. Hydrol. **19**: 154-164.

Orial G., Castanier S., Le Métayer G. and Loubiere J.-F. 1993. The Biomineralization: A new Process to Protect Calcareous Stone Applied to Historic Monuments. *In* Ktoishi, H., T. Arai and K. Kenjo (eds.), Proceed. 2nd Intern. Conf. on Biodeterioration of Cultural Property, Yokohama, Japan, **2**: 98-116.

Perthuisot, J.-P., Castanier, S., Le Métayer-Levrel, G. and Loubière, J.F. 1997. From bacteria to crystals in karstic waters. The role of nutritional conditions. *In* IAS-ASF Workshop on microbial mediation in carbonate diagenesis, Chichilianne, France, Publ. ASF, Paris **26**: 55-56.

Phillipon, J., Jeanette, D. and Lefèvre, R.-A. 1992. La conservation de la pierre monumentale en France. CNRS, Paris.

Rautureau, M. 1991. Tendre comme la Pierre. University of Orléans, France.

Riege, H., Gerdes, G. and Krumbein W.E. 1991. Contribution of heterotrophic bacteria to the formation of $CaCO_3$-aggregates in hypersaline microbial mats. Kieler Meeresforsch. Sonderh. **8**: 168-172.

Verrecchia, E.P. and Loisy, C. 1997. Carbonate precipitation by fungi in terrestrial sediments and soils. *In* IAS-ASF Workshop on microbial mediation in carbonate diagenesis, Chichilianne, France, Publ. ASF, Paris, **26**: 73-74.

BACTERIAL GENES INVOLVED IN CALCITE CRYSTAL PRECIPITATION

Brunella Perito[1], Lucia Biagiotti[1], Simona Daly[1], Alessandro Galizzi[2], Piero Tiano[3] and Giorgio Mastromei[1]
[1]Department of Animal Biology and Genetics "Leo Pardi", University of Florence, via Romana 17, I-50125 Florence, Italy. [2]Department of Genetics and Microbiology "A. Buzzati-Traverso", University of Pavia, via Abbiategrasso 207, I-27100 Pavia, Italy. [3]CNR, C.S. "Opere d'Arte", via degli Alfani 74, I-50121 Florence, Italy

Key words: calcite crystals, *Bacillus subtilis*, calcite precipitation genes

Abstract: The natural precipitation of calcium carbonate crystals by bacteria has been proposed for conservative interventions in monument restoration, even if the use of heterotrophic viable organisms does not always seem appropriate for this purpose. In fact, chemical reactions with stone minerals due to metabolic by-products and the growth of fungi, due to the application of organic nutrients for bacterial development, can have negative effects on stone surfaces. We have studied crystal precipitation in a laboratory strain of *Bacillus subtilis*. The effectiveness of this process has been analyzed on small bioclastic limestone samples (Pietra di Lecce), determining their total porosity, amount of water absorbed and superficial cohesion. To overcome some of the possible problems involved with the use of viable bacterial cells, we have investigated the genetic mechanisms that control calcite precipitation. For this reason, we have selected *Bacillus subtilis* mutants that do not precipitate or which form calcite crystals faster. The data obtained suggest the presence of several genes involved in crystal formation.

1. INTRODUCTION

Monumental stone decay is a consequence of the weathering action of several physical, chemical and biological factors, which induce a progressive dissolution of the mineral matrix. In the case of calcareous stones the material, due to calcite leaching, increases in time its porosity and decreases its mechanical characteristics (Amoroso and Fassina, 1983).

Attempts to slow down the deterioration of monuments have been continuously made by the application of conservative treatments with inorganic or organic products (Torraca, 1976). The use of the latter presents some drawbacks due to their chemical composition and thermal expansion coefficient which are quite different from that of the stone (Camaiti et al., 1988). Besides, they are usually applied in solvent at very low concentration (2-8%) and therefore a high amount of organic solvent is dispersed in the environment. Furthermore, the treatments are usually made on monumental stones that are exposed to heavy polluted atmosphere. For this reason and considering the chemical nature of the products their efficiency in time is short and, in some cases, they can have a detrimental effect for the conservation of the stone itself.

In the last years an old method, based on the application of lime-water, was proposed and experimented on deteriorated calcareous stones in order to impart a slight water-repellence and consolidating effect (Moncrieff and Hempel 1970, Price 1984). This method creates a white thin surface applicable only on white stones (like marble) and the new calcite has no bond with the substrate.

A new approach to conservative treatment of calcareous stone was attempted by inducing calcite precipitation, inside a sample of limestone, with a mineralisation process induced by the "organic matrix" macromolecules extracted from the shell of a mollusc (*Mytilus californianus*) (Tiano, 1995). The results obtained show a good efficiency of the bioinduced calcite precipitation process, with a decrease in the amount of water absorbed and an increase of the superficial strength of the treated stone (Tiano 1995).

Another way to obtain calcite precipitation inside the stone was suggested by Orial et al. (1993) with the application of living cultures of selected calcinogenic bacterial strains.

The formation of minerals by microorganisms is a very diffuse phenomenon and many products and processes are represented which, during the last 600 billions of years, have contributed to mould the earth (Lowestam, 1981). The involvement of bacteria ranges from biologically-induced to biologically-controlled processes (Mann, 1989). Microbial precipitation of $CaCO_3$ is a common process in soil, freshwater and marine sediments. The carbonate formation can occur in different ways both as passive (Krumbein, 1972) or active bioprecipitation (Castanier et al., 1990; Rivadeneyra et al., 1991). The studies made in this field have pointed out the complexity of the phenomenon that can be influenced by the environmental physico-chemical conditions and is correlated both to the metabolic activity and the cell wall structure of microorganisms (Beveridge, 1989; Fortin et al., 1997).

Bacterial activity can control the crystallogenesis of carbonates through the environmental conditions, or carbonate crystals can start to build up on the bacterial cell wall. In a few steps a kind of cocoon is formed and, subsequently, biomineral assemblage appears to turn into true crystals, either well shaped or poorly organised, which can enclose bacterial cells within the mineral structure (Castanier et al., 1990). The type of crystals formed (vaterite, aragonite or calcite) depends both on its growing features and the bacterial strain; different bacteria precipitate different types of $CaCO_3$, and the most common crystalline forms are either spherical or polyedric. The available data suggest that bacteria play an active role in $CaCO_3$ precipitation and that mineral formation is not an indirect consequence of environmental changes, produced by the metabolic activity of bacteria (Rivadeneyra et al., 1991; Fortin et al., 1997).

2. EXPERIMENTAL RESULTS

A way to obtain precipitation of new calcite crystals inside stone pores is to apply living cultures of selected calcinogenic bacterial strains (Orial et al., 1993). However, the application of viable bacteria, inside a monumental stone, does not always seem appropriate for this field of intervention. In fact, chemical reactions with stone minerals due to metabolic by-products and the growth of fungi, due to the application of organic nutrients for bacterial development, can have negative effects on stone itself. For this reason it is important to understand how bacteria form calcite crystals to find a process for inducing calcite precipitation in the absence of viable cells.

To investigate the process of calcite crystal formation we decided to work with *Bacillus subtilis*, a well characterized laboratory microorganism. The crystals produced by the *B. subtilis* strain PB19 were shown, by Spectrophotometer FT-IR and X-ray diffraction analysis, to be composed of calcite (Tiano et al., 1999). Stone samples treated with living cells showed a reduction in water uptake of about 60% (Table 1). Reference stones, supplemented with growth medium but without bacteria, showed a biological growth due to development of airborne contaminants and a noticeable decrease in water uptake. For this reason the decrease in the amount of absorbed water measured in the treated samples must be attributed, for at least 50%, to the fact that the stone porosity was physically obstructed by the presence of a consistent layer of biological mat.

Table 1. Amount of water absorbed, after 20' of contact, by stone samples before and after incubation with *B. subtilis* PB19.

Sample treated with	Before (g/cm^2)	After (g/cm^2)	Difference (%)
PB19 + medium	6.3 ± 0.7	2.7 ± 0.5	-58
Medium only	6.6 ± 0.4	4.9 ± 0.5	-26
Water	5.0 ± 0.3	5.1 ± 0.3	2

To identify the genes involved in crystal formation we isolated *B. subtilis* mutants unable to form precipitates. *B. subtilis* PB19 was mutagenized with UV radiation and a mutant unable to form precipitates was isolated. This mutation could be transferred by transformation into another *B. subtilis* strain, showing that it affects one gene only.

Mutations obtained by UV radiation are easy to obtain, but difficult to characterize. Therefore we used another genetic approach to attempt a preliminary study of the genes involved in calcite crystal precipitation. For this purpose, we screened 1,190 mutants of *B. subtilis* 168 obtained by insertional mutagenesis in order to study the functions of the uncharacterized open reading frames identified in the sequenced *B. subtilis* genome (Kunst et al., 1997). These strains were constructed by using the pMUTIN4 plasmid, belonging to the pMUTIN vector set (Vagner et al., 1998). These vectors are currently being used for systematic inactivation of genes, without known function, by the *B. subtilis* European consortium, a network of 18 European laboratories that share a common approach to inactivate about 1200 genes. A gene-by-gene inactivation procedure (replacement strategy) has been chosen; in this case every mutated strain carries a single localized mutation obtained by destruction of a known sequence. The main properties of the pMUTIN vectors are (Fig. 1): (1) the ability to replicate in *Escherichia coli* (ColE1 replication sequences and Ap^R gene as ampicillin resistance selection marker) and the inability to replicate in *B. subtilis*, which allows insertional mutagenesis and inactivation of the target gene; (2) an erythromycin resistance (Em^R) gene, which allows the selection of *B. subtilis* transformants; (3) the reporter *lac*Z gene, which becomes transcriptionally fused to the target gene, facilitating the measurement of its expression; (4) the *lacI* gene modified to be constitutively expressed in *B. subtilis*; (5) the *Pspac* promoter (Yansura and Henner, 1984) induced by IPTG and repressed by LacI, which allows the controlled expression of genes downstream the target gene. Since most of the genes in the *B. subtilis* genome are organized in operons (Vagner et al., 1998), this last requirement is important because the addition of IPTG can remove the potential failure of the transcription of the downstream genes generated by vector integration. Such polar effects are not desirable when studying a unique gene function. Moreover it makes it possible to create conditional mutants, where the target gene expression is regulated by the inducible promoter. (6) A λ terminator of transcription, t_o, is present at the end of the Em^R gene and should prevent RNA polymerase not loaded at *Pspac* to transcribe downstream genes.

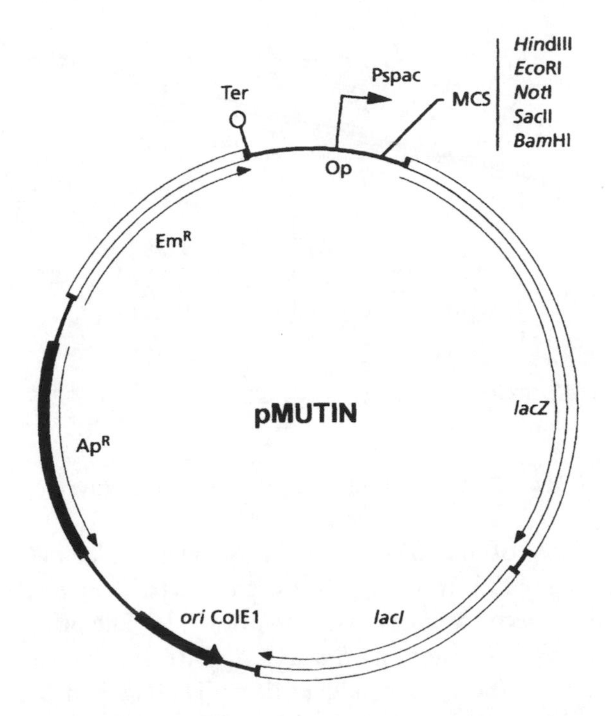

Figure 1. Map of the pMUTIN vector

The overall mutagenizing procedure consists in the amplification of the recombinant pMUTIN plasmids in *E. coli* and their use to transform *B. subtilis*. In the *B. subtilis* genome, genes can be inactivated by inserting the pMUTIN within a gene via a single crossing-over event (Fig. 2). Upon integration, the target gene is interrupted and a transcriptional fusion is generated between its promoter and the reporter *lac*Z gene. In this way two types of mutants are simultaneously generated: a first one null (orf2 in Fig. 2) via gene interruption, and a second conditional (orf3) via IPTG dependence.

The pMUTIN4 plasmid used presents additional features: a multicloning site (MCS) between the *Pspac* and the *lac*Z gene; the "oid" operator (Lehming et al., 1987) substituting the O1 operator to increase the Pspac repression by LacI repressor; two strong terminators (t_1t_2 from *rrn*B) added close to λt_o to not influence the *Pspac* activity in IPTG absence.

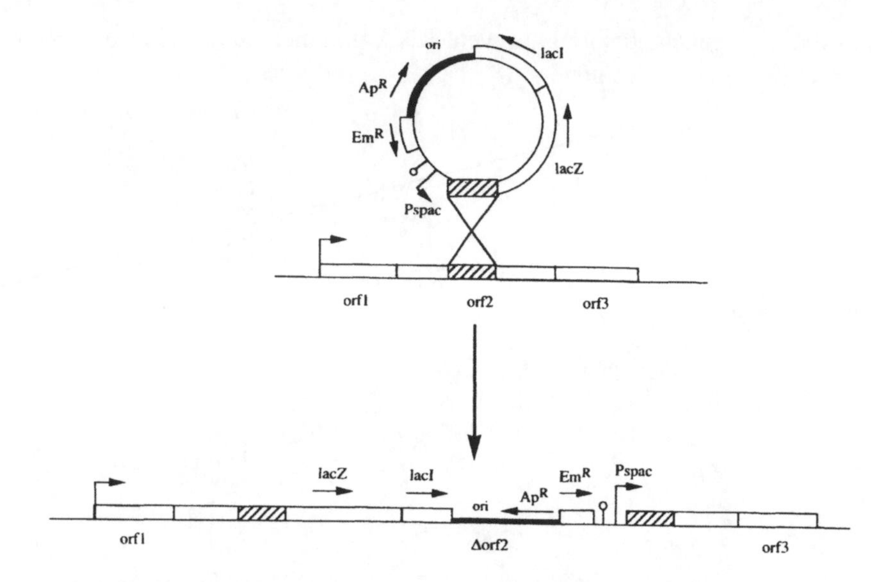

Figure 2. Integration of pMUTIN into a target gene.

The construction of the pMUTIN4 recombinant plasmids and the *B. subtilis* mutants was performed as follows: an internal fragment of the target gene was amplified from chromosomal DNA by PCR. The primers used in the reaction had an *Hind*III and a *Bam*H1 tail, respectively, to facilitate subsequent cloning of the PCR products in pMUTIN4 and the recombinant plasmid were used to transform *E. coli*. Recombinant plasmids were extracted from *E. coli* Ap^R transformants and used to transform competent cells of *B. subtilis* 168. Transformants were selected on agar plates containing 0.3 μg/ml erythromycin and analysed by Southern blot hybridisation or PCR to ensure that a single copy of the plasmid had been correctly integrated into the chromosome.

All the *B. subtilis* mutants obtained were plated on B4 medium (Rivadeneyra et al., 1991) supplemented with erythromycin (0.3 μg/ml), with and without 1 mM IPTG. Plates were incubated at 37°C and monitored for calcite precipitates production for sixteen days, comparing their behaviour to that of the *B. subtilis* PB19 wild type strain. Two groups of mutants emerged: one formed by strains which did not precipitate crystals, the other by strains which precipitated crystals faster than the control strain.

Six strains unable to form calcite crystals were isolated. Their denomination, the corresponding interrupted gene and their behaviour respect to calcite precipitation are shown in Tab. 2.

Table 2. B. subtilis mutants which do not form calcite crystals.

Strain	Interrupted gene	Calcite precipitates after days –IPTG	+IPTG
BFA1345	*yusJ*	n.d.	n.d.
BFA1346	*yusK*	n.d.	n.d.
BFA1347	*yusL*	n.d.	n.d.
BFA1847	*ykoY*	n.d.	n.d
BFA2426	*ysiA*	15	7
BFA2427	*ysiB*	15	7
PB19 (wt)		7	7

n.d.= not detectable

Four strains did not form crystals at all, neither in the absence nor in the presence of IPTG. Three of these, BFA1345, BFA1346 and BFA1347, have as interrupted genes *yusJ*, *yusK* and *yusL*, respectively. These genes are close each other, transcribed in the same direction and close to other *yus* genes (region 3366083-3372083 of the *B. subtilis* chromosome). The strain BFA1847 has the *ykoY* gene interrupted, which is grouped together with other *yko* genes (region 1409119-1411119). For these four cases, a strong involvement of the interrupted genes in calcite precipitation should be supposed, since IPTG addition does not change the phenotype. The sequence analysis of the putative products was performed by BLAST in Entrez database and has given the following indications (Tab. 3). The *yusJ* product is similar to the butyryl-CoA dehydrogenases of *Thermoanaerobacterium thermosaccharolyticum* (GI: 1903328) and to other acyl-CoA dehydrogenases like those of *Bos taurus* (GI: 1351839), *Homo sapiens* (GI: 1703068) and *B. subtilis* (GI: 2634849). The *yusK* product is similar to the acetyl-CoA C-acyltransferases and the 3-ketoacyl-CoA thiolases of *E. coli* (GI: 1070462) and *Pseudomonas fragi* (GI: 135752). The *yusL* product is similar to the 3-hydroxyacyl-CoA dehydrogenases of *Rickettsia* (GI: 3861108), *Archeoglobus fulgidus* (GI: 2650351) and *Bradyrhizobium japonicum* (GI: 3287857). All these enzymes are involved in the fatty acid beta-oxidation pathway. The *ykoY* product is similar to the toxic anion resistance proteins, such as GI: 4160468 of *Bacillus megaterium*, to the putative tellurium resistance protein of *B. subtilis* (GI: 2632578) and to other hypothetical transmembrane or integral membrane transport proteins such as GI: 1788119 of *E. coli*, GI: 2314509 of *Helicobacter pylori* and GI: 1573003 of *Haemophilus influenzae*.

Table 3. Possible functions of the genes interrupted in *B. subtilis* mutants which do not form calcite crystals.

Strain	Interrupted gene	Possible function
BFA1345	*yusJ*	Butyryl-CoA dehydrogenase*
BFA1346	*yusK*	Acetyl-CoA C-acyltransferase*
BFA1347	*yusL*	3-hydroxyacyl-CoA dehydrogenase*
BFA1847	*ykoY*	Membrane transport protein
BFA2426	*ysiA*	Transcriptional regulator (TetR/AcrR family)
BFA2427	*ysiB*	3-hydroxybutyryl-CoA dehydratase*

*fatty acid metabolism

Two strains, BFA2426 and BFA2427, formed crystals as fast as the control strain when IPTG is present, whereas in the absence of IPTG some crystal formation could be observed only after fifteen days of incubation (Tab. 2). Moreover their interrupted genes, *ysiA* and *ysiB*, are located one after the other in the same transcription direction (region 2916058-2918058, Fig. 3). In this case a possible involvement of downstream genes in calcite production should be expected, since IPTG addition, which promotes the expression of genes downstream of the target gene, seems to restore the wild type behaviour. The *ysiA* product is similar to prokaryotic transcriptional regulators belonging to the TetR/AcrR family: among these, the regulatory protein MTRR of *Neisseria gonorrhoeae* (GI: 730078), which controls the permeability of the cell envelope to hydrophobic compounds. The *ysiB* product is similar to the 3-hydroxybutyryl-CoA dehydratases involved in fatty acid metabolism: among these there are enzymes similar to the *yusL* product and the *yusL* product itself. The genes immediately downstream of *ysiA* and *ysiB* in the same transcription direction are *etfB*, *etfA*, whose products (GI: 2635318 and GI: 2635317) are similar to the electron transfer flavoproteins beta and alpha subunits respectively.

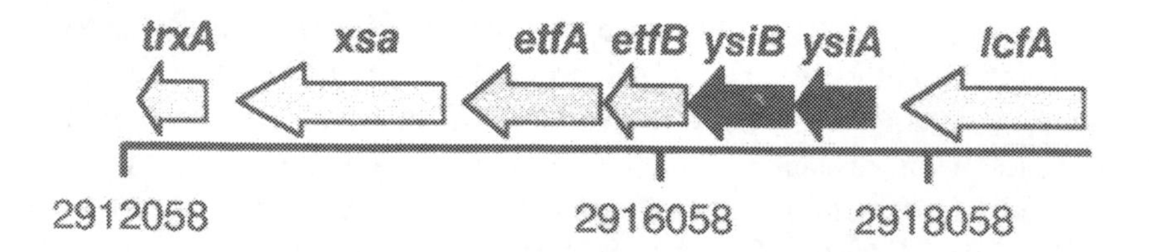

Figure 3. Map of the *B. subtilis* chromosome region containing the *ysiB* and *ysiA* genes. Arrows show the direction of transcription.

The group of mutants which form calcite crystals faster is formed by six strains (Tab. 4). In every case IPTG addition seems to restore the wild type phenotype. This effect is more visible with the fastest strains (BFA1303, BFA1308 and BFA3002), suggesting the possible involvement not only of the interrupted gene but also of the following gene(s) in the same transcription direction. In fact, restoring the expression of downstream gene(s) could eliminate the deregulation due to the interruption of the upstream target gene.

Table 4. B. subtilis mutants which form calcite crystals faster.

Strain	Interrupted gene	Calcite precipitates after days –IPTG	+IPTG
BFA1303	*ywqH*	2	7
BFA1304	*ywqj*	5	7
BFA1308	*ywqB*	2	6
BFA2232	*ywoE*	3	6
BFA3002	*yitD*	2	6
BFA3005	*yitG*	5	7
PB19 (wt)		7	7

Strains BFA1303, BFA1304 and BFA1308 carry, as interrupted genes, *ywqH*, *ywqJ* and *ywqB*, respectively (Tab. 3). These genes are grouped with other *ywq* genes (region 3723753-3735753). *YwqB* is transcribed on one strand, whereas *ywqH* and *ywqJ* are transcribed on the other strand; *ywqH* preceding, in order, *ywqI* and *ywqJ*. The precipitation speed was greater in the case of *ywqH*, suggesting the involvement of downstream genes, such as *ywqI* and *ywqJ*. A similar speed was found for *ywqB*, with a possible involvement of the *ywqA* downstream gene. The sequence analysis of the putative products gave the following indications (Tab. 5). The putative product of *ywqH* is similar to proteins, from various organisms, which, beyond their specific functions, share the common feature to be aggregation substances. Some examples are: surface aggregation proteins involved in sex pheromone-inducible conjugation in *Enterococcus faecalis* such as Asp1 (GI: 282273), Asc10 (GI: 97895) and GI: 114245; the PfEMP1 membrane protein of *Plasmodium falciparum* (GI: 2645461), the putative adhesin product of *Enterococcus faecium ash* gene (GI: 2109266). The *ywqJ* product is similar to the YxiD protein and other hypothetical proteins from *B. subtilis* with unknown function. The *ywqB* product showed no obvious similarity with known proteins. The putative product of the downstream *ywqA* gene showed high homology to helicases of the SNF2/RAD54 family. Strain BFA2232 has the *ywoE* interrupted gene which is close to other upstream and downstream *ywo* genes, very near to the *ywq* genes (region about 3752333). The sequence analysis has shown that the *ywoE* product is similar to transmembrane transport proteins like the uracil permease of *Schizosaccharomyces pombe* (GI: 2492816), the putative allantoin permease of *E. coli* (GI: 2498121) and the hypothetical *Streptomyces coelicolor* transmembrane transporter GI: 3127838.

The *yitD* and *yitG* genes are interrupted in strains BFA3002 and BFA3005, respectively. They lie in opposite transcription direction, close to other upstream and downstream *yit* genes (region 1171806-1177806). The *yitD* product is similar to hypothetical proteins which could share a common feature of adhesion substances, such as the putative fibronectin/fibrinogen–binding protein of *Helicobacter pylori* (GI: 2314561) and a cadherin similar protein of *Caenorhabditis elegans* (GI: 3193145). The BFA3002 mutant strain carrying the interrupted *yitD* gene showed, together with BFA1303 and BFA1308, the fastest precipitation speed. On the other hand, *yitD* is followed by six *yit* genes oriented in opposite transcription direction, making it difficult to hypothesise an involvement of downstream genes on the mutant phenotype. The *yitG* product is similar to multidrug resistance proteins, by a transmembrane efflux mechanism, belonging to the major facilitator family also known as the drug resistance translocase family. Among these, the tetracycline resistance protein class C of *E. coli* (GI: 135549), the putative multidrug-efflux transporter YfmO of *B. subtilis* (GI: 2633053), the MffT2 transporter of *Aquifex aeolicus* (GI: 2983666). It should be noted that, among these resistance proteins homologous to *yitG* product, there is the *ywoG* product; *ywoG* being a *ywo* gene near to *ywoE* but in opposite transcription orientation.

Table 5. Possible functions of the genes interrupted in *B. subtilis* mutants which form calcite crystals faster.

Strain	Interrupted gene	Possible function
BFA1303	*ywqH*	Aggregation protein
BFA1304	*ywqJ*	Unknown
BFA1308	*ywqB*	Unknown
BFA2232	*ywoE*	Transmembrane permease
BFA3002	*yitD*	Adhesion protein
BFA3005	*yitG*	Drug resistance translocase

3. CONCLUSIONS AND PERSPECTIVES

The application of *B. subtilis* cells to stone samples induces a lowering in stone porosity, as verified with the capillary water absorption tests, but this phenomenon seems mostly due to pore physical obstruction. Moreover, we must consider that this treatment can have negative consequences, such as: (a) deposition of new products, due to chemical reactions between the stone minerals and by-products originating from the bacterial metabolism; (b) stained patches, due to the growth of air-borne fungi and related to the presence of organic nutrients, necessary for bacterial development. To overcome these drawbacks and to improve this approach to monumental stones conservation, the genes controlling the mineral forming process should be identified in order to obtain calcite precipitation under better controlled conditions.

The results reported in this work represent the first data available on the genes involved in calcite precipitation. The most remarkable information emerging from the sequence analysis of the mutated genes, involved in crystal formation, is the great number of translational products which, beyond their specific functions, share the common feature to be membrane associated proteins. This is the case for the putative products of *ykoY*, *ywqH*, *ywoE*, and *yitG*. Among these, the *ywqH* product could be also an adhesion factor, like as the *yitD* product. Membrane associated proteins are also the flavoproteins coded by *etfB* and *etfA*, the genes downstream *ysiA* and *ysiB*, and probably involved in the calcite phenotype of the BFA2426 and BFA2427 strains. The other prevalent feature seems to be the involvement of enzymes with functions in the fatty acid metabolism: the putative products of *yusJ*, *yusK*, *yusL* and *ysiB*. The calcite phenotype of the strains carrying the *yus* genes (BFA1345, BFA1346 and BFA1347) is not affected by IPTG addition, suggesting that only the interrupted genes are involved.

The isolation of mutants which form calcite crystals faster than the parental strain offers the possibility to speed up the process and, therefore, reduces the problems described before.

AKNOWLEDGMENTS

We thank Sandro Costa for the screening of the mutant strains. This work was partially supported by the EU contract BIO4-CD95-0278.

REFERENCES

Amoroso, G. and V. Fassina. 1983. Material Science Monographs 11, Elsevier, Amsterdam.

Beveridge, T.J. 1989. Role of cellular design in bacterial metal accumulation and mineralization.. Annu. Rev. Microbiol. **43**: **147-171**.

Camaiti, M., G. Borselli and U. Matteoli. 1988. Prodotti consolidanti impiegati nelle operazioni di restauro. Edilizia **10**: 125-134.

Castanier, S., A. Maurin and J.P. Perthuisot. 1990. A trial to get dolomite in freshwater. Geobios **23:** 121-128.

Fortin, D., F.G. Ferris and T.J. Beveridge. 1997. Surface-mediated mineral development by bacteria. Rev. Mineral. **35**: 161-180.

Krumbein, W.E. 1972 Roles des microorganismes dans la génèse la diagénèse et la dégradation des roches en place. Rev. Ecol. Biol. Sol. **3:** 283-319.

Kunst, F., N. Ogasawara, I. Moszer and 148 other authors. 1997. The complete genome sequence of the gram-positive bacterium *Bacillus subtilis*. Nature **390:** 249-256.

Lehming, N., J. Sartorius, M. Niemoller, G. Genenger, B. Wilcken-Bergmann and B. Muller-Hill. 1987. The interaction of the recognition helix of *lac* repressor with *lac* operator. EMBO J. **6:** 3145-3153.

Lowestam, H., A. 1981. Minerals formed by organisms. Science **211**: 1126-1131.

Mann, S. 1989. Crystallochemical strategies in biomineralization. *In* Mann, S., J. Webb and R.J.P. Williams (eds.), Biomineralization chemical and biochemical perspectives, VCH, Germany, p. 35-62.

Moncrieff, A. and K. Hempel. 1970. Work on the degeneration of sculptured stone. Conservation of stone and wooden objects, New York Conference, 2nd Edition, ICC Ed., London, **1**: 103-114.

Orial, G., S Castanier, G. Le Metayer and J.F. Loubiere. 1993. The biomineralization: a new process to protect calcareous stone applied to historic monuments. *In* Ktoishi, H., T. Arai and K. Kenjo (eds.), Proceeding of 2nd International Conference on Biodeterioration of Cultural Property, Yamano publisher, Yokohama, Japan, **2**: 98-116.

Price, C. A. 1984. The consolidation of limestone using a lime poultice and lime-water. Adhesives and Consolidants, Paris Meeting, ICC Ed., London p. 160-162.

Rivadeneyra, M.A., R. Delgado, E: Quesada and A. Ramos-Cormenzana. 1991. Precipitation of calcium carbonate by *Deleya halophila* in media containing NaCl as sole salt. Curr. Microbiol. **22:** 185-190.

Tiano, P. 1995. Stone reinforcement by calcite crystals precipitation induced by organic matrix macromolecules. Studies in Conservation **40:** 171-176.

Tiano P., L. Biagiotti and G. Mastromei. 1999. Bacterial bio-mediated calcite precipitation for monumental stones conservation: methods of evaluation. J. Microbiol. Methods **36**: 139-145.

Torraca, G. 1976. Treatment of stone in monuments – a review of principles and processes. *In* The Conservation of Stone I, R. Rossi Manaresi (ed.), Bologna, Italy p. 297-315.

Vagner, V., E. Dervyn, and S.D. Ehrlich. 1998. A vector for systematic gene inactivation in *Bacillus subtilis*. Microbiology **144:** 3097-3104.

Yansura, D.G., and D.J. Henner. 1984. Use of the *Escherichia coli lac* repressor and operator to control gene expression in *Bacillus subtilis*. Proc.Natl. Acad. Sci. USA. **81:** 439-443.

BIOREMEDIATION OF CULTURAL HERITAGE: REMOVAL OF SULPHATES, NITRATES AND ORGANIC SUBSTANCES

Giancarlo Ranalli[1], Mauro Matteini[2], Isetta Tosini[2], Elisabetta Zanardini[3] and Claudia Sorlini[3]

[1]*Department of Food and Microbiological Sciences and Technologies, University of Molise, via De Sanctis 46, I-86100 Campobasso, Italy;* [2]*Opificio Pietre Dure, Fortezza da Basso, I-50129 Firenze, Italy;* [3] *Department of Food and Microbiological Sciences and Technologies, University of Milan, via Celoria 2, I-20133 Milano, Italy.*

Key words: bioremediation of artworks, nitrates, sulphates, organic matter

Abstract: Among the different pathologies of stone materials, the presence of sulphates and nitrates is the most frequent. In addition, the presence of organic matter on artistic stoneworks can be attributed to inadequate past restorations, to the lysis of microbial cells of the primary surface colonisation, and to the presence of hydrocarbons originating from oil combustion. The latter appears to be a serious danger for the preservation of artworks themselves. Until today, chemical and physical techniques have been largely used to remove pollutants and residual substances from works of art by using surfactants and solubilizing agents. We developed a multiple bioremediation system for the biological removal of sulphates, nitrates and organic matter present on artistic stoneworks utilizing microbial cultures carefully selected and grown on a suitable support. The study consisted of: screening of microorganisms in order to select cultures with a high sulphate-reducing, denitrifying and biodegradative ability of organic matter; setting up of simulated laboratory tests with stone samples artificially enriched with nitrates, sulphates and organic matter; testing appropriate inert matrices on which to immobilize the selected bacterial strains; and testing sulphate, nitrate and organic matter removal from artificially enriched stones as well as from naturally degraded artworks. Sepiolite was used to develop bacterial biofilms with a high active biomass per cm^3. However, in order to eliminate or reduce interferences and the release of undesirable ions and contaminant trace elements, Hydrobiogel-97, a mixture of polyacrylamide at different molecular weights, appears to be a good carrier. As regards nitrate removal, *Pseudomonas aeruginosa* and *P. stutzeri* were selected for their high denitrifying activity. Treatment with colonised sepiolite of stone specimens artificially enriched with nitrates and of real stone samples showed that, after 30 h, a very high percentage (90% and 88%, respectively) of nitrate was removed. For sulphate removal, *Desulfovibrio vulgaris* and *D. desulfuricans* were selected and tested in liquid cultures, on stone specimens artificially enriched with sulphates, and on real marble samples. The highest removal efficiency (81%) was achieved on real marble samples after 36 h of treatment. Finally, for the removal of organic matter, bioremediation tests on ancient frescoes (XV century) located near Pisa, altered by the presence of undesirable residual collagen, were carried out before intervention with traditional restoration. The treatment with pure cultures of *P. stutzeri*, at a temperature of 17-22°C, showed the complete removal of collagen after 8 h. The results confirm the potentiality of bioremediation processes as soft innovative technology based on the use of microorganisms and their metabolic activity in the recovery of degraded artworks.

1. STATE OF THE ART

Among the pathologies present on different outdoor artistic works, those related to lithoid materials (stones, frescoes, paintings) are the most well known. In fact, in the last decades, particularly in urban areas, alterations such as black crusts, nitratation, sulphation and, in particular conditions, the presence of dust, residual hydrocarbons and other organic pollutants have markedly increased, as shown by the rapid diffusion of phenomena of corrosion. All the studies confirm that the alterations are caused by inorganic atmospheric pollutants such as nitrogen oxides and sulphur dioxide produced by the combustion of petroleum and its derivatives. The compounds, which are the main cause of accelerated deterioration of exposed stoneworks, are oxidized in the air into nitric and sulphuric acid, respectively, that, after deposition on the surface of stone carbonates, are converted into sulphates and highly soluble nitrates, which are easily washed away by rain.

With regard to the presence of nitrates, they can originate from the reaction of numerous oxides of nitrogen present, in percentage of parts per million, in the polluted atmosphere (N_2O, NO, N_2O_3, NO_2, N_2O_5), through the process of oxidation and reaction with the water vapour, to produce as final products nitrous acid and more abundant nitric acid (HNO_3) (1). Such acids contribute to the formation of acid rain and attack metal and stone sculptures located outdoors, as well as paintings, causing the transformation of insoluble calcium carbonate into a soluble salt of calcium nitrate (2):

$$NOx + H_2O \rightarrow HNO_3 \quad (1)$$

$$CaCO_3 + 2HNO_3 \rightarrow Ca(NO_3)_2 + H_2O + CO_2 \quad (2)$$

Sulphates can originate from the oxidation of sulphur dioxide (annual production in the world of SO_2 in the nineties is estimated at about 400 million tons or more) into sulphur trioxide due to the presence of oxygen (3). sulphur trioxide reacts in the atmosphere with water to form sulphuric acid (H_2SO_4) (4). The fraction of sulphuric acid that is not neutralised in the air contributes to the formation of acid rain. The steady increase in the last few years of the acidity values of rain water represents a potential risk for humans, nature and artworks by causing the transformation of insoluble calcium carbonate into more soluble calcium sulphate and carbon dioxide (5).

$$2SO_2 + O_2 \rightarrow 2SO_3 \quad (3)$$
$$SO_3 + H_2O \rightarrow H_2SO_4 \quad (4)$$
$$CaCO_3 + H_2SO_4 \rightarrow CaSO_4.2H_2O + CO_2 \quad (5)$$

Another kind of alteration are black crusts, which are composed of crystals of gypsum (including recrystallized calcite) mixed with atmospheric particles (spores, pollen, dust and different particulate matter called "smog", heavy hydrocarbons like those of the aromatic and aliphatic series) entrapped in the mineral matrix. It is commonly believed that they originate from wet and dry deposition processes in which sulphuric acid, a sulphur dioxide oxidation product, attacks carbonic rocks, resulting in gypsum formation (Saiz-Jimenez, 1991; Saiz-Jimenez and Garcia, 1991).

Finally, the presence of organic matter on artistic stoneworks can be ascribed to inadequate past restoration, to the lysis of microbial cells of primary surface colonisation, and to the presence of hydrocarbons originating from oil combustion. This fact appears to represent a serious danger for the preservation of the artworks. The phenomena are particularly evident when the artistic stoneworks are located in the open air, where atmospheric pollution can contribute to accelerate the degradation of materials. Among the environmental factors that influence such processes there is humidity, wind, rainfall, thermal variations, and organic and inorganic pollutants.

The pathologies of the materials composing the works of art are generally cured by chemical and/or physical techniques, whereas biological methods are scarcely used. In fact, only some enzymes have been applied to the recovery of some alterations. In contrast, whole living microorganisms as agents of biorecovery have never been utilized; only sporadic studies under laboratory conditions have been carried out.

As regards the possible removal of the deposition of nitrates on frescoes by biological treatment, only preliminary information has been reported by Gabrielli (1981). The author indicated the possibility of using denitrifying microorganisms and the importance to perform preliminary experimental tests in order to identify the optimal conditions (pH, cultural media and anaerobic conditions). The biological process of nitrate removal is based on the use of a wide range of facultative anaerobic bacteria that are able to utilise nitrates as terminal respiratory electron acceptors. Denitrifying bacteria, such as *Pseudomonas* sp., are present in different natural environments (wet soil, water and wastewater). In anaerobic conditions, they are able to reduce nitrates to gaseous products like nitrous oxide and molecular nitrogen, which at room temperature evolve into the atmosphere (6) (Knowles, 1982).

$$NO^{3-} \rightarrow NO^{2-} \rightarrow N_2O \rightarrow N_2 \quad (6)$$

In the aforementioned studies, the experiments were carried out using sulphate-reducing bacteria as broth cultures on altered stone samples. The technique uses strictly anaerobic bacteria capable of reducing sulphate to hydrogen sulphide, which in turn evolves into the atmosphere. In natural habitats, bacteria belonging to genera of *Desulfovibrio, Desulfomonas, Desulfobacter, Desulfococcus,* and *Desulfosarcina* are present in anaerobic mud and freshwater sediments, in marine environments, and in the gastrointestinal tract of man and animals (Postgate, 1984).

The removal of sulphates has been considered by Atlas and Rude (1988). Aliquots of broth containing the sulphate-reducing bacterium *Desulfovibrio desulfuricans* were used to treat marble samples showing black weathering crusts rich in gypsum. The results demonstrated that calcite was found on all treated surfaces, suggesting the potential use of this microorganism to clean encrusted marble. The authors suggested that the formation of crystals occurs by a combination of dissolution-precipitation and diffusion processes; calcium ions (Ca^{2+}), released from gypsum when the bacteria reduce sulphates, react with carbon dioxide and result in the formation of calcite, according to the following equation (7):

$$6CaSO_4 + 4H_2O + 6CO_2 \rightarrow 6CaCO_3 + 4H_2S + 2S + 11\,O_2 \quad (7)$$

However, the authors concluded that further studies should be carried out to fully understand the mechanism of microbial calcification in order to determine the crystallographic continuity of the neomineralized calcite, as well as to try to isolate the enzymes responsible for sulphate reduction.

Heselmeyer et al. (1991) studied in laboratory conditions the possibility to obtain the bioconversion of rock gypsum crusts into calcite. They reported that the conventional physico-chemical cleaning methods are often disadvantageous, because they can cause colour changes in the rocks or excessively remove the original rock material. For the removal of sulphates, a strain of *Desulfovibrio vulgaris* was used, and the treatment was performed on gypsum-encrusted marble and on sandstone in anaerobic conditions. At the end of the experiment, no sulphate was detectable in the marble sample, whereas in the sandstone a reduction of about 40% was observed.

In order to determine the rate and extent of sulphate depletion from the solution by a strain of *Desulfovibrio desulfuricans*, Gauri and Gwinn (1983) performed laboratory experiments on marble specimens altered by exposure to 10 ppm of SO_2 at 100% relative humidity and 20°C. The capability of sulphate reduction by the strain in liquid culture was more than 80% after 60 h. In the preliminary experimental phase, it is important to optimise the cultural medium in order to avoid the production of a black FeS precipitate. With regard to the treatment on stone samples immersed in the cultural broth, the authors demonstrated the production of H_2S. Further investigations were carried out by treating for 24 h with the same strain sections of marble statues which showed black regions determined by particles of atmospheric soot and dust cemented by gypsum The utilisation of the microorganism did not clean the surfaces and the marble samples remained brownish. However, the authors suggested that regulation of the process and of the physiological conditions (anaerobiosis and temperature) could be improved in order to favour bacterial growth and activity.

Other studies have evaluated the possibility of using microorganisms involved in the precipitation of calcium carbonate (Boquet et al., 1973), independently of sulphate reduction: the salt is present in natural environments (soil, fresh water and saline habitats) such as marine calcareous skeletons, carbonate sediments, soil carbonates and carbonate rocks. The exact role of bacteria in precipitation is still not fully understood (Krumbein, 1979). In fact, active and passive roles have been proposed for bacteria including the possibility to act as crystal nuclei and to determine calcium concentration and bicarbonate ion production (Erlich, 1981). Krumbein (1979) and Krumbein and Giele (1979) extensively studied carbonate deposition caused by growth and the metabolic activity of photosynthetic microorganisms (cyanobacteria) associated with the formation of desert stromatolites, which are organosedimentary structures produced by sediment trapping and/or precipitation. Rivadeneyra et al. (1985) noted that some environmental characteristics could affect the precipitation of $CaCO_3$, such as the ionic strength of the medium (2.5-20% total salt concentration). For this reason, moderately halophytic bacteria isolated from sea water assume a relevant importance in the study of the physiological process. Calcium carbonate precipitation is influenced by salinity, temperature and viscosity, and the latter in particular interferes with the rate of the process and the crystal characteristics.

In a subsequent work, Rivadeneyra et al. (1994), using 63 isolated strains belonging to the genus *Vibrio*, showed, in all the tested conditions, the formation of crystals of magnesium calcite, with a variable Mg content, depending on the medium provided. In any case, no aragonite was detected even in a medium with a high Mg content.

The experimental trials showed that Mg^{2+} has an inhibitory effect on carbonate precipitation by bacteria, and the optimal temperature for the highest number of crystal-forming strains was 32°C instead of 22°C. However, such findings held true only when the external conditions were not ideal for precipitation (Rivadeneyra et al., 1991). Another factor that can influence the regulation of the crystallisation process is biopolymer production, such as extracellular polysaccharides. In particular, Kok et al. (1986) observed in a marine alga, *Emiliana huxleyi*, the probable involvement of a polysaccharidic complex in the crystallisation process (of Ca or Mg) of calcite inside intracellular vacuoles; they suggested that the mechanism of agglomeration was stimulated through viscous binding between two crystal surfaces.

Recently, Tiano et al. (1999) reported a new approach for inducing a bio-mediated precipitation of calcite directly inside the stone porosity through the application of organic matrix macromolecules extracted from sea shells or of living calcinogenic bacteria. Laboratory tests were developed in order to evaluate parameters such as porosity, superficial strength and chromatic changes. The authors indicated that this type of treatment might not be suitable for monumental stone conservation.

In this paper are reported the results of our research carried out in order to develop and improve a multiple bioremediation system for removal of nitrates, sulphates and organic matters from altered artistic stoneworks; the process consists of the utilisation of microbial cultures carefully selected, on the use of a suitable cell support and on its direct application on altered surfaces.

2. RECENT LABORATORY INVESTIGATIONS ON BIOREMEDIATION OF ARTISTIC STONEWOKS

Before submitting deteriorated artworks to bioremediation processes, investigations were performed following four different steps: i) screening of microorganisms with a high sulphate-reducing activity, denitrifying activity and high biodegrading ability of specific organic matter; ii) setting up of simulated laboratory tests using selected cultures on stone samples artificially enriched with nitrates, sulphates and organic matter; iii) selection of an optimal inert matrix to use as a carrier of microbial cells; iv) improving the removal of sulphates, nitrates and organic matter from artificially enriched stones (Ranalli et al., 1996, 1997).

2.1 Microorganisms and biofilm selection

Microorganisms for the bioremediation of artworks can be obtained from international cultural collections (bacteria, yeasts, moulds) and by isolation from different natural habitats (soil, wastewaters, cultural heritage samples, etc.) on the basis of their specific metabolic activities. They can be used in pure culture, in co-cultures, and in mixed cultures. Many factors such as temperature, relative humidity, pH, aerobic or anaerobic conditions, carbon and energy sources, etc. can influence the metabolic activities. They must be accurately controlled in laboratory tests in order to optimise the metabolic efficiency. The conceptual premise of the tests can be summarized as follows: if bioremediation cannot be achieved under optimal conditions (at laboratory scale), then it is highly unlikely that it can be achieved in any other non-controlled biological technology. In order to avoid failures, it is thus necessary to verify whether the laboratory results can be obtained even under real environmental conditions, particularly for outdoor artworks.

Treatment based on the bioremediation of artworks can be performed by applying the selected microorganisms to the surface by spray, brush or compress. The choice of the different forms of application depends on the type of alteration, the artwork material, the location of the areas to be submitted to the treatment, and the metabolic activity of the selected microflora (aerobic and anaerobic). The time course of the treatment can be evaluated in order to define the effectiveness of the biological process. If a long duration of the treatment, high concentration of biomass, a favourable contact between microorganisms and surface are required, it is necessary to consider the possibility to use an adequate support as a cell carrier.

The selection of the best support should be based on its ability to offer adequate hydration without interference with the substrate by release of ions or with the microbial metabolic processes. Furthermore, the matrix should be easy to remove and not toxic for the operators or the environment. Preliminary laboratory tests demonstrated that sepiolite (Tulsa, Spain) was the best inert material to develop microbial biofilms with a high quantity of active biomass per cm^3 (Ranalli et al., 1995). However, before direct use, the support must be submitted to a pretreatment process to eliminate or reduce interference and the release of undesirable ions and traces of contaminant (Ranalli et al., 1996). In some tests performed on samples of Lecce stone artificially enriched with sulphates, the presence of very small amounts of iron on the lithotype caused irregular, large brownish-black spots as a consequence of re-precipitation as iron sulphides (FeS).

Recently, different supports represented by physico-chemical non-filmogen polymeric gels, composed of a mixture of functionalised acrylamide (PAAM) with different molecular weights, as well as of poly-vinyl alcohol, were tested as possible carriers for microbial activity. "Hydrobiogel-97" was the most suitable for our treatment conditions of lithoid materials such as different types of marble and stone. Under our experimental conditions, the biogel allowed optimal hydration even after nine days of treatment; it also did not show any chromatic, chemical or physical modification in its composition or on the stone material (data not shown).

The biological approach used for bioremediation on stoneworks is depicted in Fig.1.

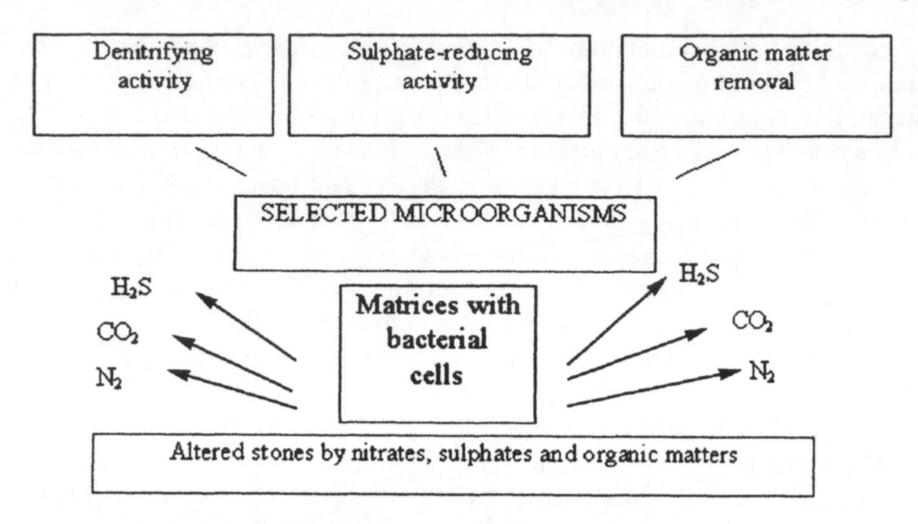

Figure 1. Scheme of biological approach used for bioremediation of stoneworks.

2.2 Bioremediation with denitrifying bacteria

To evaluate the possibility to biologically remove nitrates from stone materials, it is necessary during the first phase of experimental trials, to select the microbial cultures in defined media and to test them on stone samples artificially enriched with nitrates. However, the risk of using artificially altered samples is the rapid formation on the stone surface of thin layers of crystallised and highly soluble salts. In such cases, the efficiency of the treatment may be strongly influenced by the loss of nitrate even in the control sample without microorganisms.

For nitrate removal, pure cultures of *Pseudomonas aeruginosa* (RZ94) and *Pseudomonas stutzeri* (GB94) were selected for their denitrifying activity in different natural environments and compared with other culture collection strains. In preliminary tests, the influence of different nitrate salts on biological removal was analysed on brick specimens covered with a double layer of enriched plaster with Terra d'Ombra and Sinopia Red. The enrichment of plaster was obtained by the addition of calcium nitrate [$Ca(NO_3)_2$], sodium nitrate [$NaNO_3$], potassium nitrate [KNO_3] and barium nitrate [$Ba(NO_3)_2$]. For the treatment, a broth culture of *Pseudomonas stutzeri* strain (GB94) was used at two different cell concentrations (10^5 and 10^7 cells/ml). The specimens, after 45 days of drying at room temperature, showed a white efflorescence in the presence of sodium and potassium nitrate, whereas no efflorescence was observed with the other salts. The controls were performed by applying sterile water instead of the broth cultures. During three days of treatment, the specimens were covered by a plastic film in order to favour a partial anaerobic condition and to maintain a high relative humidity.

The best results in terms of biological nitrate removal were obtained at high cell concentrations for calcium nitrate, on the surface and deep within the lithoid samples (34% average value). A lower removal efficiency was noted in the case of Na, K and Ba nitrates.

Numerous subsequent laboratory tests with sepiolite activated by *Pseudomonas stutzeri* (GB94) were performed on stone specimens artificially enriched with nitrates (initial nitrate contents ranged in weight between 3.0% and 6.0%) and on real stone samples (initial nitrate content 1,100 ppm). Nitrates present in a very high percentage (average values from 88% to 90%) were removed after 30 h at 28°C in anaerobic conditions (Ranalli et al., 1996).

Fig. 2 shows the effect of biological removal of nitrate on a real sample of Vicenza stone after 30 h of treatment with sepiolite activated by *Pseudomonas stutzeri*.

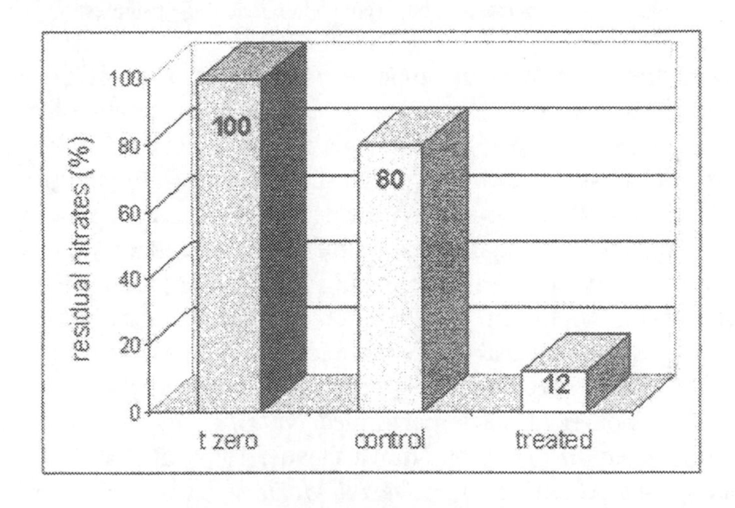

Figure 2. Biological nitrate removal on real stone samples.

2.3 Bioremediation with sulphate-reducing bacteria

For the sulphate-reducing bacteria, *Desulfovibrio vulgaris* and *D.desulfuricans* were selected for sulphate removal. To evaluate their activity, treatment was performed with colonised sepiolite on artificially enriched stone (initial content 1.0%, 2.3% and 3.0% w/w) and on real *ex-situ* marble samples (initial content 1,900 and 1,350 ppm). After the treatment, the highest removal efficiency was achieved on real marble (81% after 36 h) (Ranalli et al., 1997) as reported in Fig. 3.

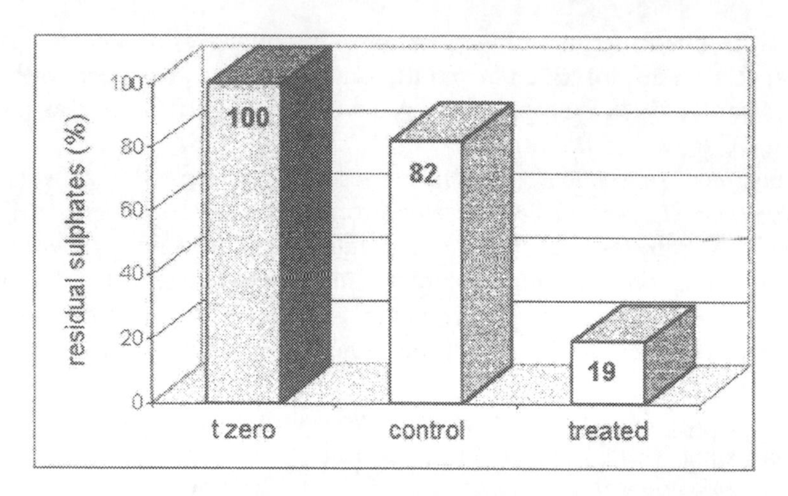

Figure 3. Biological sulphate removal on real stone samples.

However, from the data it appears important to consider numerous environmental factors that can directly influence the metabolism of the selected microflora in order to optimize the efficiency of the biological removal of sulphates. For example, when the temperature of the process, the treatment times or cultural conditions are not well defined (like an inadequate buffer solution), some damage to the native crystallographic structure of the artwork material may occur. The following SEM observations (Fig. 4) show the effects of the biological removal of sulphate and possible damage: corrosion of the stone material as a consequence of an acidification of the cell suspensions after a too prolonged treatment is evident in Fig. 4c.

Subsequent biological tests performed *in situ* on altered stoneworks exposed to the open air (Taverna Nuova façade, 1574, at Popoli, Italy) were carried out using a gel activated by *Desulfovibrio vulgaris* (10^7 cells/ml). The selected areas (surface about 400 cm^2 for treatment and control without microorganisms at 1.4% w/w initial sulphate concentration), after application with the colonized Hydrobiogel-97, were covered using an appropriate synthetic film that permits the maintenance of humidity and anaerobic conditions required for the metabolic biological processes. The *in situ* experiment was performed at ambient temperature (about 6-8°C during the winter season) for 7 days. The results showed a promising value of sulphate removal (average 28%), considering the restricted environmental conditions of the microbial activities adopted.

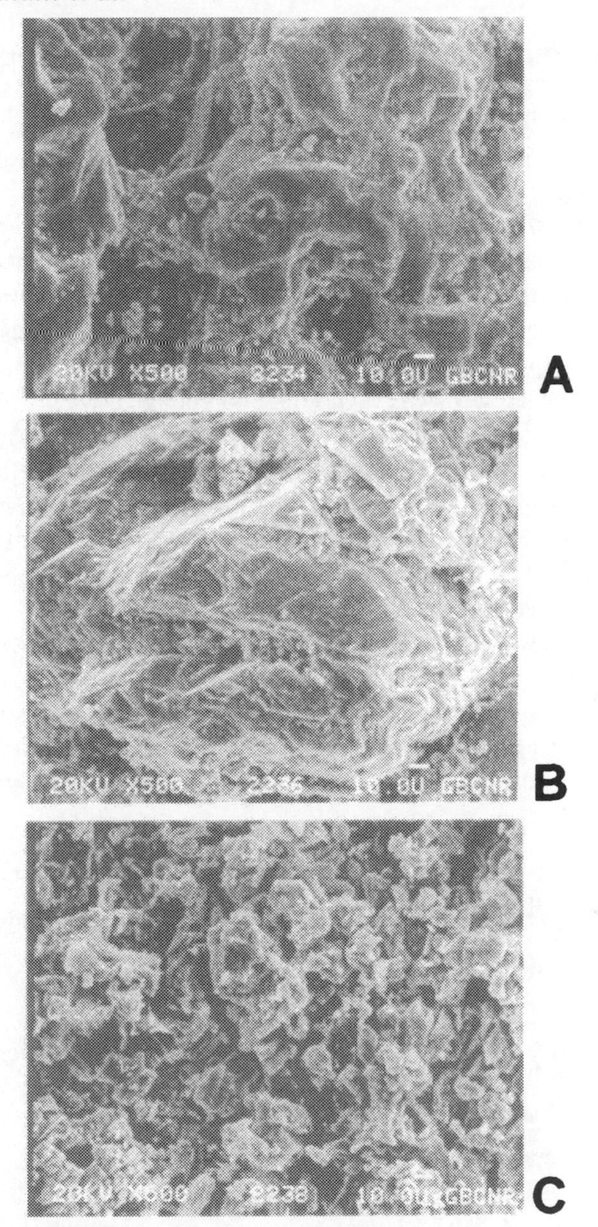

Figure 4. SEM observation: A) black crust on altered marble surface; B) marble surface after appropriate biological sulphate removal; C) marble surface after a too prolonged treatment.

2.4 Biological removal of organic matter

With regards to organic matter removal, recent bioremediation tests were recently carried out on ancient frescoes (XV century), painted by Spinello

Aretino and Andrea Bonaiuti and located at Camposanto Monumentale (Pisa), altered by a high content of undesirable residue of collagen added in the past as a glue. The collagen, widely distributed on the fresco surface, represented a major problem for restorers during the traditional intervention of restoration. In fact, it showed a very low solubility in water solutions with or without the addition of specific enzymes, even at different ambient temperatures. However, a semi-mechanical removal approach is inadequate because it takes too long and results in some damage. For such reasons, pure cultures of *Pseudomonas stutzeri* strain A29 were selected on the basis of the high biodegradative activity on different organic substrates, such as casein, walnut oil, linseed oil, shellac, beeswax, egg yolk and collagen, when supplied as sole carbon and energy sources under aerobic conditions. Different tests showed a marked influence on the temperature of biological treatments. In fact, after a daily application of cell suspensions (about 10^8 CFU/ml) on altered fresco surfaces (treated areas about 900-76,000 cm^2) at environmental temperatures ranging from 17 to 22°C, an initial removal of collagen was observed by the third day and it was completed by the fifth day, compared with the control areas (Fig. 5). However, no evident biological effects were noted when the room temperature was maintained below 10°C. On the basis of the preliminary positive results, the Technical Commission for Restoration (Pisa) has accepted and approved the utilisation of *Pseudomonas stutzeri* in pure cultures, and the bioremediation process is still being extensively used.

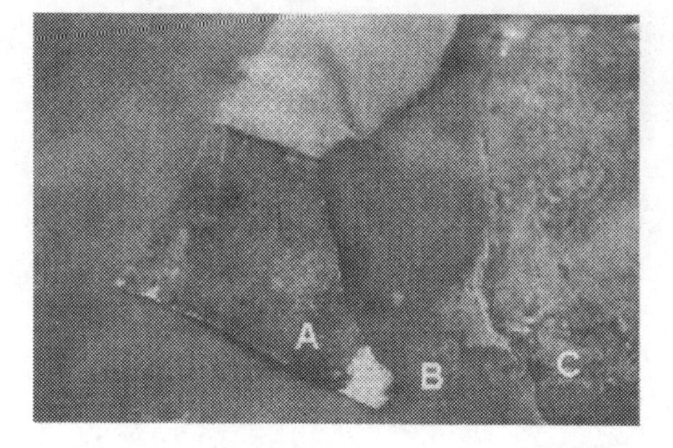

Figure 5. A typical fresco surface covered with a cloth and collagen glue (A); area after mechanical removal (B); and a fragment after biological treatment with selected bacteria (C).

3. MONITORING OF THE BIOLOGICAL PROCESSES OF BIOREMEDIATION ON CULTURAL HERITAGE

In all cases of bioremediation in which microorganisms are used for artwork restoration, a final cleaning phase must be carried out. In order to prevent undesired metabolic processes from continuing, residual microflora should be carefully removed at the end of the specific biological process. Prior to commencing any bioremediation process, an appropriate strategy must be considered and defined with respect to the protection of the artworks. When microorganisms are not controlled in time, they may still be living and active and therefore become, directly or as a consequence of the effects of the by-products of their catabolism, the causes of material loss or damage. Among the several possibilities to remove microflora, it is suggested a preliminary soft mechanical removal of rough residual material (such as inorganic matrix used as carrier of microbial cells) followed by a gentle cleaning with demineralised or, even better, sterile water. Nevertheless, when the peculiarities of the artistic sample submitted to the bioremediation by microorganisms permits other interventions, appropriate surfactant solutions and weak biocides can be used.

In order to identify the optimal conditions to develop a bioremediation approach for artworks by using selected microflora, one of the most important aspects is the monitoring of microbial activity during and at the end of the biological process. Among the different techniques, microbial viable counts and direct counts by microscope observation are routinely carried out. SEM and optical microscope observations offer important information on the presence, the grade of adhesion between microflora and materials, the interaction with the microflora already present on the altered artwork, and the dynamics of the microbial community. New and rapid bioindicators, like ATP content and DHA (dehydrogenase activity), can be utilised to monitor microbial activity.

4. CONCLUSIONS

Nitrates, sulphates and organic matter (such as collagen, beeswax, casein, walnut oil, etc.) present on the surface of artworks can be removed, in appropriate conditions, by treating with selected microbial cultures of denitrifying, sulphate-reducing and aerobic heterotrophic bacteria. The biological process, as a consequence of activities of the selected cultures, releases nitrogen, H_2S and carbon dioxide from the tested pollutants without the use of toxic compounds and, consequently, must be considered an environmentally compatible method.

After the evaluation and optimisation of the parameters of the biological process on a laboratory scale and subsequent control of the results of the treatment on outdoor samples, further application in other fields requires the setting up of several requisites: i) a continuous and abundant activated biomass production on appropriate matrices in pilot plants under strictly controlled environmental conditions; ii) the use of rapid and sensitive methods to monitor the biological activities and to avoid undesired effects on the surface materials; iii) a correct and complete removal of residual biological activities. Studies are in progress to apply in situ the capabilities of bioremediation methods on artistic stoneworks and to reduce their limitations.

In conclusion, the results herein reported confirm the potentiality of bioremediation processes as soft innovative technology based on the use of microorganisms and their metabolic activity for the recovery of degraded artworks.

ACKNOWLEDGMENTS

The study was developed under the National Research Program on Chemicals, second phase, supported by the *Ministero dell'Università e Ricerca Scientifica e Tecnologica* (MURST) and assigned to Syremont S.p.A., Milano (Italy).

The authors thank G. Pizzigoni (Syremont S.p.A.) and the restorer G. Caponi (Pisa) for their friendly co-operation. They gratefully acknowledge L. Tinucci, P. Albonico and M. Castellano (EniTecnologie S.p.A., San Donato) for providing the physico-chemical gel.

REFERENCES

Atlas, R.C. and P.D. Rude. 1988. Complete oxidation of solid phase sulfides by manganese and bacteria in anoxic marine sediment. Geochimica et Cosmochimica Acta **52**: 751-766.

Boquet, E., A. Boronat and A. Ramos-Cormenzana. 1973. Production of calcite (calcium carbonate) crystals by soil bacteria is a general phenomenon. Nature **246**: 527-529.

Erlich, H.L. 1981. Geomicrobiology. Marcell Dekker, New York.

Ferrer, M.R., J. Quevedo-Sarmiento, M.A. Rivadeneira, V. Bejar, R. Delgadoand and A. Ramos-Cormenzana. 1988. Calcium carbonate precipitation by two groups of moderate halophilic microorganisms at different temperatures and salt concentrations. Curr. Microbiol. **17**: 221-227.

Gabrielli, N. 1981. Proposte di ricerca. Bollettino dei Monumenti Musei e Gallerie Pontificie: pp. 123-124.

Gauri, K. L. and J. A. Gwinn. 1983. Deterioration of marble in air containing 5-10 ppm SO_2 and NO_2. Durability of Building Materials **1**: 217-233.

Heselmeyer, K., U. Fischer, W.E. Krumbein and T. Warsheid. 1991. Application of *Desulfovibrio vulgaris* for the bioconversion of rock gypsum crusts into calcite. BIOforum **1/2**: 89.

Knowles, R. 1982. Denitrification. Microbiol. Rev. **46**: 43-70.

Kok, D.J., L.J. Blomen, P. Westbroek and O.L. Bijvoet. 1986. Polysaccharide from coccoliths ($CaCO_3$ biomineral). Influence on crystallization of calcium oxalate monohydrate. Eur. J. Biochem. **158**: 167-172.

Krumbein, W.E. 1979. Phototropic and chemoorganotropic activity of bacteria and algae as related to beachrock formation and degradation (Gulf of Agaba, Sinai). Geomicrobiol. J. **1**: 139-203.

Krumbein, W.E. and C. Giele, 1979. Calcification in a coccoid cyanobacterium associated with the formation of desert stromatolites. Sedimentology **26**: 593-604.

Postgate, J.R. 1984. Genus *Desulfovibrio* Kluyver and van Niel 1936. *In* Bergey's Manual of Systematic Bacteriology, N.R. Krieg and J.G. Holt (eds.), Williams and Wilkins, Baltimore/London p. 666-672.

Ranalli G., M. Chiavarini, V. Guidetti, F. Marsala, M. Matteini, E. Zanardini and C. Sorlini. 1996. The use of microorganisms for the removal of nitrates and organic substances on artistic stoneworks. *In* Proceedings of 8th International Congress on Deterioration and Conservation of Stone, Berlin (Germany) p. 1421-1427.

Ranalli G., M. Chiavarini, V. Guidetti, F. Marsala, M. Matteini, E. Zanardini and C. Sorlini. 1997. The use of microorganisms for the removal of sulphates on artistic stoneworks. Int. Biodet. Biodegr. **40**: 255-261.

Rehr, B. and J.H. Klemme. 1989. Competition for nitrate between denitrifying *Pseudomonas stutzeri* and nitrate ammonifying enterobacteria. FEMS Microbiol. Ecol. **62**: 51-58.

Rivadeneyra, M.A., R. Delgado, A. del Moral, M.R. Ferrer and A. Ramos-Cormenzana. 1994. Precipitation of calcium carbonate by *Vibrio* spp. from an inland saltern. FEMS Microbiol. Ecol. **13**: 197-204.

Rivadeneyra, M.A., R. Delgado, E. Quesada and A. Ramos-Cormenzana. 1991. Precipitation of calcium carbonate by *Deleya* halophila in media containing NaCl as sole salt. Curr. Microbiol. **22**: 185-190.

Rivadeneyra, M.A., I. Perez-Garcìa, V. Salmeròn and A. Ramos-Cormenzana. 1985. Bacterial precipitation of calcium carbonate in presence of phosphate. Soil Biol. Biochem. **17**: 171-172

Saiz-Jimenez, C. 1993. Deposition of airborne organic pollutants on historic buildings. Atmosp. Environ. **27B**: 77-85.

Saiz-Jimenez, C. 1995. Exposure of anthropogenic compounds on monuments and their effect on airborne microorganisms. Aerobiology **11**: 161-175.

Saiz-Jimenez C. and M.A. Garcia del Cura. 1991. Sulfated crusts: a microscopic, inorganic and organic analyses. *In* Science, Technology and European Cultural Heritage, N.S. Baer, C. Sabbioni and A.I. Sors (eds.), Butterworth-Heinemann, London p. 523-526.

Tiano, P., L. Biagiotti and G. Mastromei. 1999. Bacterial bio-mediated calcite precipitation for monumental stones conservation: methods of evaluation. J. Microbiol. Methods **36**: 139-145.

Index